T0350273

Design and Optimization of Sensors and Antennas for Wearable Devices

Vinod Kumar Singh
S. R. Group of Institutions Jhansi, India

Ratnesh Tiwari
Bhilai Institute of Technology, India

Vikas Dubey
Bhilai Institute of Technology, India

Zakir Ali
IET Bundelkhand University, India

Ashutosh Kumar Singh
Indian Institute of Information Technology, India

A volume in the Advances in Mechatronics and Mechanical Engineering (AMME) Book Series

Published in the United States of America by
IGI Global
Engineering Science Reference (an imprint of IGI Global)
701 E. Chocolate Avenue
Hershey PA, USA 17033
Tel: 717-533-8845
Fax: 717-533-8661
E-mail: cust@igi-global.com
Web site: http://www.igi-global.com

Library of Congress Cataloging-in-Publication Data

Names: Singh, Vinod Kumar, 1977- editor. | Tiwari, Ratnesh, 1982- editor. | Dubey, Vikas, 1985- editor. | Ali, Zakir (Electronics authority), editor. | Singh, Ashutosh Kumar, editor.
Title: Design and optimization of sensors and antennas for wearable devices / Vinod Kumar Singh, Ratnesh Tiwari, Vikas Dubey, Zakir Ali, and Ashutosh Kumar Singh, editors.
Description: Hershey, PA : Engineering Science Reference, an imprint of IGI Global, [2020] | Includes bibliographical references.
Identifiers: LCCN 2019006763| ISBN 9781522596837 (hardcover) | ISBN 9781522596844 (softcover) | ISBN 9781522596851 (ebook)
Subjects: LCSH: Wearable technology--Antennas. | Detectors.
Classification: LCC TK7871.6 .D434 2020 | DDC 621.39/167--dc23 LC record available at https://lccn.loc.gov/2019006763

This book is published in the IGI Global book series Advances in Mechatronics and Mechanical Engineering (AMME) (ISSN: 2328-8205; eISSN: 2328-823X)

British Cataloguing in Publication Data
A Cataloguing in Publication record for this book is available from the British Library.

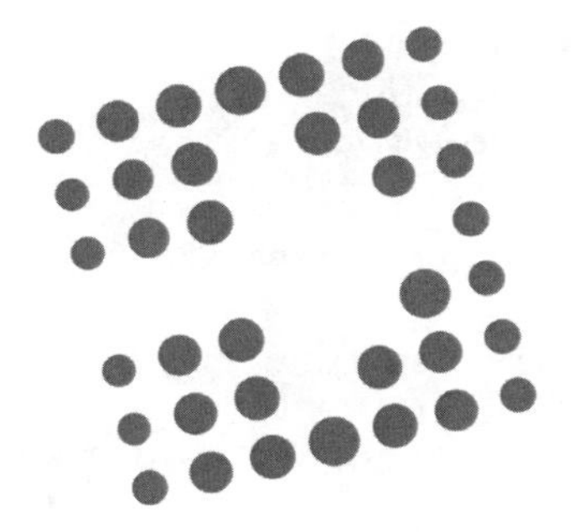

Advances in Mechatronics and Mechanical Engineering (AMME) Book Series

ISSN:2328-8205
EISSN:2328-823X

Editor-in-Chief: J. Paulo Davim, University of Aveiro, Portugal

MISSION

With its aid in the creation of smartphones, cars, medical imaging devices, and manufacturing tools, the mechatronics engineering field is in high demand. Mechatronics aims to combine the principles of mechanical, computer, and electrical engineering together to bridge the gap of communication between the different disciplines.

The **Advances in Mechatronics and Mechanical Engineering (AMME) Book Series** provides innovative research and practical developments in the field of mechatronics and mechanical engineering. This series covers a wide variety of application areas in electrical engineering, mechanical engineering, computer and software engineering; essential for academics, practitioners, researchers, and industry leaders.

COVERAGE

- Vibration and acoustics
- Control Methodologies
- Control Systems Modelling and Analysis
- Micro and nanomechanics
- Intelligent Navigation
- Medical robotics
- Sustainable and green manufacturing
- Computational Mechanics
- Intelligent Sensing
- Nanomaterials and nanomanufacturing

IGI Global is currently accepting manuscripts for publication within this series. To submit a proposal for a volume in this series, please contact our Acquisition Editors at Acquisitions@igi-global.com or visit: http://www.igi-global.com/publish/.

The Advances in Mechatronics and Mechanical Engineering (AMME) Book Series (ISSN 2328-8205) is published by IGI Global, 701 E. Chocolate Avenue, Hershey, PA 17033-1240, USA, www.igi-global.com. This series is composed of titles available for purchase individually; each title is edited to be contextually exclusive from any other title within the series. For pricing and ordering information please visit http://www.igi-global.com/book-series/advances-mechatronics-mechanical-engineering/73808. Postmaster: Send all address changes to above address.

Titles in this Series

For a list of additional titles in this series, please visit:
http://www.igi-global.com/book-series/advances-mechatronics-mechanical-engineering/73808

Global Advancements in Connected and Intelligent Mobility Emerging Research and Opportunities
Fatma Outay (Zayed University, UAE) Ansar-Ul-Haque Yasar (Hasselt University, Belgium) and Elhadi Shakshuki (Acadia Universit, Canada)
Engineering Science Reference • ©2020 • 278pp • H/C (ISBN: 9781522590194) • US $195.00

Automated Systems in the Aviation and Aerospace Industries
Tetiana Shmelova (National Aviation University, Ukraine) Yuliya Sikirda (Kirovograd Flight Academy of the National Aviation University, Ukraine) Nina Rizun (Gdansk University of Technology, Poland) Dmytro Kucherov (National Aviation University, Ukraine) and Konstantin Dergachov (National Aerospace University – Kharkiv Aviation Institute Ukraine)
Engineering Science Reference • ©2019 • 486pp • H/C (ISBN: 9781522577096) • US $265.00

Design, Development, and Optimization of Bio-Mechatronic Engineering Products
Kaushik Kumar (Birla Institute of Technology, India) and J. Paulo Davim (University of Aveiro, Portugal)
Engineering Science Reference • ©2019 • 312pp • H/C (ISBN: 9781522582359) • US $225.00

Nonlinear Equations for Problem Solving in Mechanical Engineering
Seyed Amir Hossein Kiaeian Moosavi (Babol Noshirvani University of Technology, Iran) and Davood Domairry Ganji (Babol Noshirvani University of Technology, Iran)
Engineering Science Reference • ©2019 • 250pp • H/C (ISBN: 9781522581222) • US $195.00

For an entire list of titles in this series, please visit:
http://www.igi-global.com/book-series/advances-mechatronics-mechanical-engineering/73808

701 East Chocolate Avenue, Hershey, PA 17033, USA
Tel: 717-533-8845 x100 • Fax: 717-533-8661
E-Mail: cust@igi-global.com • www.igi-global.com

Table of Contents

Preface

The main objective of this book is to present wearable system and flexible compact antennas for wireless sensor network and energy harvesting applications. The anticipated book is providing deep introduction about wearable body area network, which has been utilized in telemedicine, which will be definitely helpful for the society. There are several 3D full wave EM software such as IE3D, CST, ADS, and HFSS, which are utilized to design flexible wearable antennas. The flexible wearable antennas design in this book was simulated using IE3D and CST software. Jeans are flexible material that is used to make receiver and transmitter of wearable antenna, which play most important technological role for mankind.

In the present decade most of the electronic devices are consuming low power and this power is supplied by means of batteries .However these batteries are rechargeable ones but powering them is a difficult task. Even though power consumption is low only solar cells could power them, no other sources provide a great deal of energy. On the other side energy harvesting will provide the better solution to powering the wearable electronics and wireless sensors. Radio Frequency harvesting technique of energy conversion has assumed great importance during the last few decades because of excellent flexible materials have been developed.

Chapter 1 describes a 'ला' shape antenna for high frequencies is designed which has been simulated under CST Software using Copper material i.e. FR-4. The dielectric constant of this material is 4.3. The return loss of 'ला' shaped antenna is -28 dB at 6.774 Giga-Hertz and -19 dB at 7.7 Giga-Hertz resonant frequencies.

Chapter 2 describes the hammer shaped textile antenna having large bandwidth, which is used in wide band applications. The model of antenna is simulated in CST software. The optimization of anticipated antenna has been studied in this chapter.

Chapter 3 reports a hybrid wearable energy harvesting system. Integration of microwave antenna on thin film amorphous silicon solar cell creates a hybrid system which can harvest both the solar and microwave energies.

In this Chapter 4, dual wideband textile antenna is proposed for WLAN and WiMax application. Textile Antennas are invaluable as a result of their compelling cost and straightforward acknowledgment process.

Chapter 5 proposed triple band novel geometry & enhanced characteristics of flexible antenna. Textile Antenna is exceptionally invaluable due to their successful cost, small profile, low mass and basic acknowledgment process.

Chapter 6 explains the concept of energy harvesting and also its potential function in wireless system. This paper reports a hybrid wearable energy harvesting system. Integration of microwave antenna on thin film amorphous silicon solar cell creates a hybrid system which can harvest both the solar and microwave energies.

Chapter 7 describe multiband zigzag shaped microstrip patch antenna with cross Cut-set patch with defected ground structure has been designed, simulated, and tested for wireless applications.

Chapter 8 describes the concept of multilevel inverter with different conventional topologies with different pulse width modulation techniques and shows the comparison between the total harmonic distortions.

Chapter 9 describes Microstrip rhombus patch antenna, which contain many properties, includes, light weight, low profile, low cost, less volume and easy to install on rigid surface due to these properties it is easy to fabricate.

Chapter 10 describes, Slotted Wearable Antenna which is designed at frequency 2.4 GHz for wireless application and radiolocation. Proposed antenna is used for radiolocation through which detection of objects is possible using a tracking system of radio waves by analyzing the properties of received radio waves.

Chapter 11 describes the single element of wearable antenna is designed and further to enhance the gain of wearable rectenna array is designed. The anticipated antenna array shows the directivity of 8.048 dBi that was used to calculate received power by antenna array. This rectenna array can be used to operate the micro-electronic gadgets and to operate small sensors.

In Chapter 12, different textile materials have been discussed which are flexible in nature and light weight. E textile is one such wearable antenna where the sensors are integrated with the very fabric to be used by humans for various purposes and objectives Usage. of e-textiles in medical facility or at the times of disaster management etc makes it perhaps one of the most important wearable antennas because of the scale of purpose it aims to solve.

In Chapter 13 a relationship is developed between the theoretical inset feed distance and simulated coaxial feed distance of rectangular patch antenna. A ratio also has been developed between the inset feed distance and coaxial feed distance. By knowing one feed distance other feed distance can be easily determined. This relationship can help in simulation process of design a co-axial probe feed rectangular patch antenna.

As this book covers relatively wide areas and numerous contents connected to complex scientific issues, errors and omissions may be unavoidable due to the limited knowledge and competence of the authors, therefore we sincerely appreciate the criticism and comments from the readers. We feel confident it will be highly valued for its documentation of the important field of optimization of sensors and antennas for wearable device and its practical applications.

Vinod Kumar Singh
S. R. Group of Institutions Jhansi, India

Ratnesh Tiwari
Bhilai Institute of Technology, India

Vikas Dubey
Bhilai Institute of Technology, India

Zakir Ali
Bundelkhand University, India

Ashutosh Kumar Singh
Indian Institute of Information Technology Allahabad, India

Chapter 1
A "La" Shape Antenna for High Frequencies Applications

Anurag Saxena
S. R. Group of Institutions Jhansi, India

ABSTRACT

In this chapter, a 'ला' shape antenna for high frequencies is designed which has been simulated under CST software using copper material (i.e., FR-4). The dielectric constant of this material is 4.4. The return loss of 'ला' shaped antenna is -28 dB at 6.774 giga-hertz and -19 dB at 7.7 GHz resonant frequencies. It covers the bandwidth from 6.555 GHz to 7.122 GHz and 7.38 GHz to 8.07 GHz. In this chapter, simulated results like polarization, smith chart, return loss graph, 2-D pattern, 3-D pattern, and polar plot are presented.

INTRODUCTION

In the city or covered wireless environment, after a complicated multiple reflection or scattering effect, the polarization of the propagating radio wave may change significantly. Although various recent wireless systems are vertically polarized, it has been predicted that it is advantageous to use horizontally polarized antennas at transmitter and receiver ends (Saxena and Singh, 2018a; Saxena, Singh, Bhardwaj, Chae, Sharma, and Bhoi, 2018; Saxena and Singh, 2018b; Singh and Saxena, 2018c).

DOI: 10.4018/978-1-5225-9683-7.ch001

The selection of microstrip antenna technology can fulfill these requirements. The main problem is that they usually radiate in a direction along the ground plane, and the gain in the horizontal direction is only a few decibels (Singh and Saxena, 2018d; Verma, Singh, and Saxena, 2016; Toore, Vallozzi, Rogier and Verhaevert, 2010; Lemey, Delercq and Rogier, 2014; Chahat, Zhadobov and Sauleau, 2014).

The broadband antennas are required to be compact, low-profile, directive with high transmission efficiency and designed to be discreet. These antennas are usually large metallic cumbersome objects. For dual band systems directivity and security are important features. It is important for the system to be highly directive in order to reduce coverage in unwanted areas (Lui, Murphy and Toumazou, 2013; Agilent, 2005; Roundy, 2003; Shukla, Verma, Gohir, 2015; Hall and Hao, 2006; Manel, Patil and Dhanawade, 2014).

Use of fractal geometry enhances the impedance bandwidth of an antenna structure based on its self-similarity property, due to which several resonances corresponding to different iterative structures are excited, overlapping with each other to result in wide bandwidth. The space-filling property increases the electrical length of the antenna structure without increasing its physical size. Some fractal antenna structures for SWB applications have been described in literature (Muralill, Maha, Dilip and Chaitanyal, 2014; Zhang, Chai, Xioa, and Ye, 2013; Kennedy, Fink, Chu, Champange, Lin and Khayat, 2009; Chahat, Zhadobov, and Sauleau, 2014; Lui, Murphy, and Toumazou, 2013; Torre, Vallozzi, Rogier and Verheavert, 2010; Lemey, Declercq and Rogier, 2014).

DESIGN CONFIGURATION OF PRESENTED ANTENNA

The presented antenna is imprinted on FR-4 epoxy substrate having thickness of 1.6 mm in which the dielectric permittivity of this material is 4.3 and loss tangent of 0.02. It can feed by a 50 Ω rectangular microstrip line and defected partial ground plane. The dimensions were calculated using standard design equations,

$$50 = \frac{120\pi}{\sqrt{\varepsilon reff}\left[\frac{W_f}{h} + 1.393 + 0.667\ln\left(\frac{W_f}{h} + 1.444\right)\right]}$$

Figure 1. Front view and Back view of presented antenna

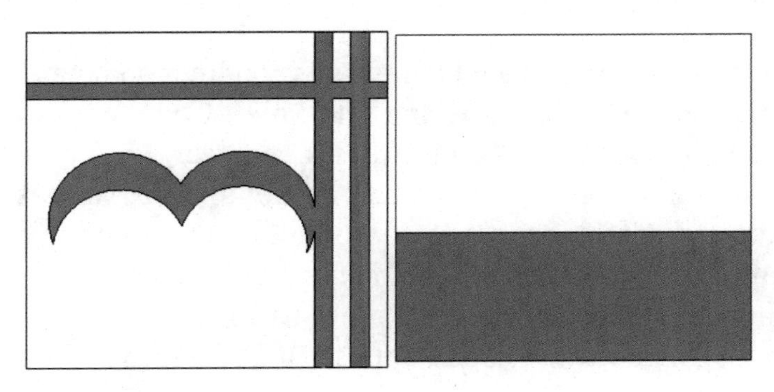

Table 1. Parameters of Presented Antenna

Parameters	Value
Relative permittivity (ε_r)	4.4
Resonant Frequency-1	6.774 GHz
Resonant Frequency-2	7.7 GHz
Substrate Dimension	50 x 50 mm
Strip Line-1	2.5 x 50 mm
Strip Line-2	2.5 x 50 mm
Height of Material	1 mm

$$\varepsilon_{reff} = \frac{\varepsilon_r + 1}{2} + \frac{\varepsilon_r - 1}{2}\left[1 + 12\frac{h}{W}\right]^{-.05}$$

The design of presented antenna having front view and back view is shown in figure 1. The dimension ground made by copper is 50 x 22 mm having 0.0038 mm height and the dimension of substrate is 50 x 50 mm. The two patch circle is designed on copper substrate with two circle cuts for making design "ल". The first circle cut shape is having radius of 9.5 mm and second circle cut shape is having radius of 9 mm. Table 1 shows the various parameters of presented antenna.

All the simulation and designing were optimized using Computer Simulation Technology (CST) software. Each antenna parameter was optimized stepwise with all other parameters fixed. In each optimization, best impedance bandwidth performance was chosen. Perfect partial ground plane and conductors were assumed.

SMITH CHART

For solving the problems related to transmission lines and matching circuits in radio frequency (RF) engineering, the Smith Chart was invented. This is a graphical aid designed for electronics engineer, Phillip H. Smith was invented the Smith Chart. With the help of Smith chart various parameters like impedances, admittances, reflection coefficients, scattering parameters, noise figure circles, constant gain contours etc can be display for unconditional stability, including mechanical vibrations analysis.

The Smith Chart of 'ला' shape antenna at high frequencies at two resonant frequencies is shown in Figure 2.

POLARIZATION

There will be polarization in presented 'ला' shape antenna at high frequencies. In antenna, polarization is of two types; left polarization and right polarization. Both polarizations are shown in fig-3, fig-4, fig-5 and fig-6 at two resonant frequencies.

The radiation efficiency of left polarization at 6.774 GHz and 7.7 GHz frequency is -1.558 dB and -2.122 dB. The radiation efficiency of right polarization at 6.77 GHz and 7.7 GHz frequency is -1.558 dB and -2.122 dB.

Figure 2. Smith Chart of Presented Antenna

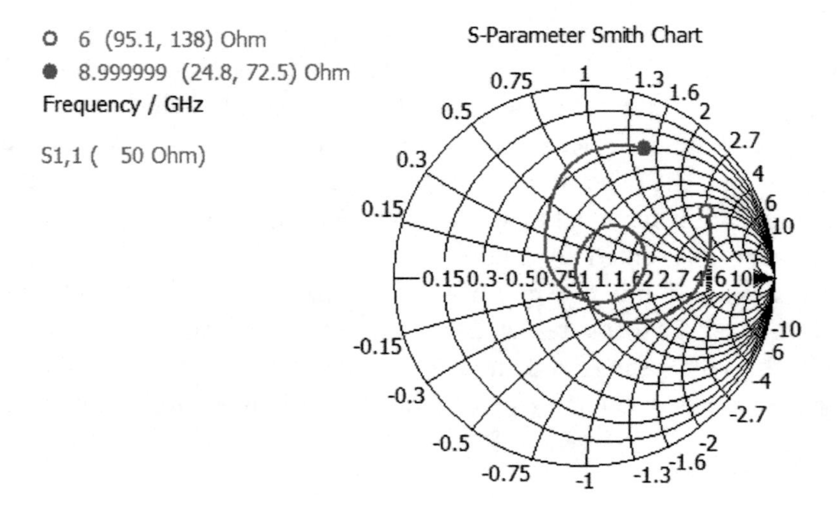

Figure 3. Left Polarization at 6.774 GHz resonant frequency

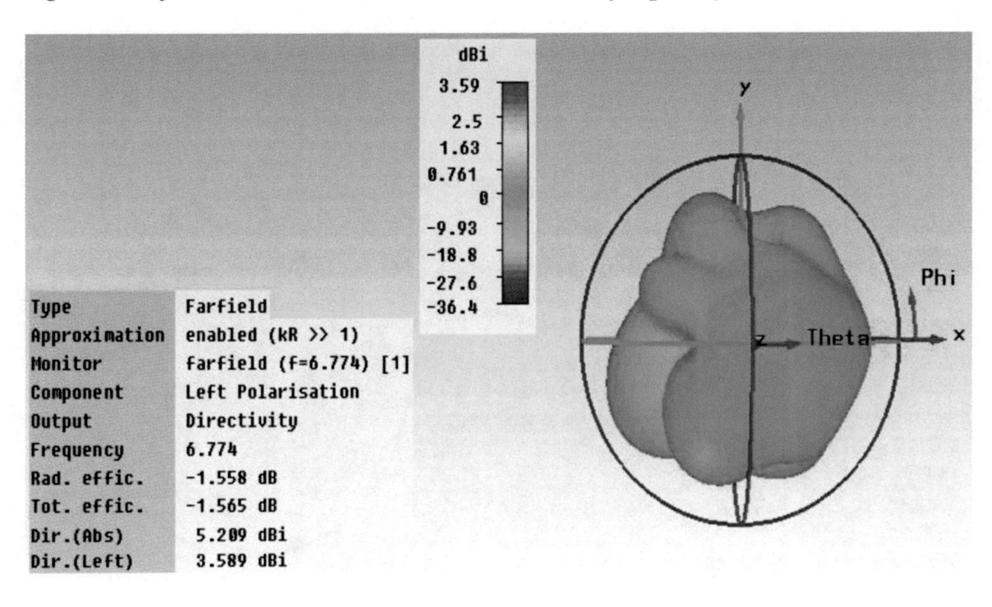

Figure 4. Right Polarization at 6.774 GHz resonant frequency

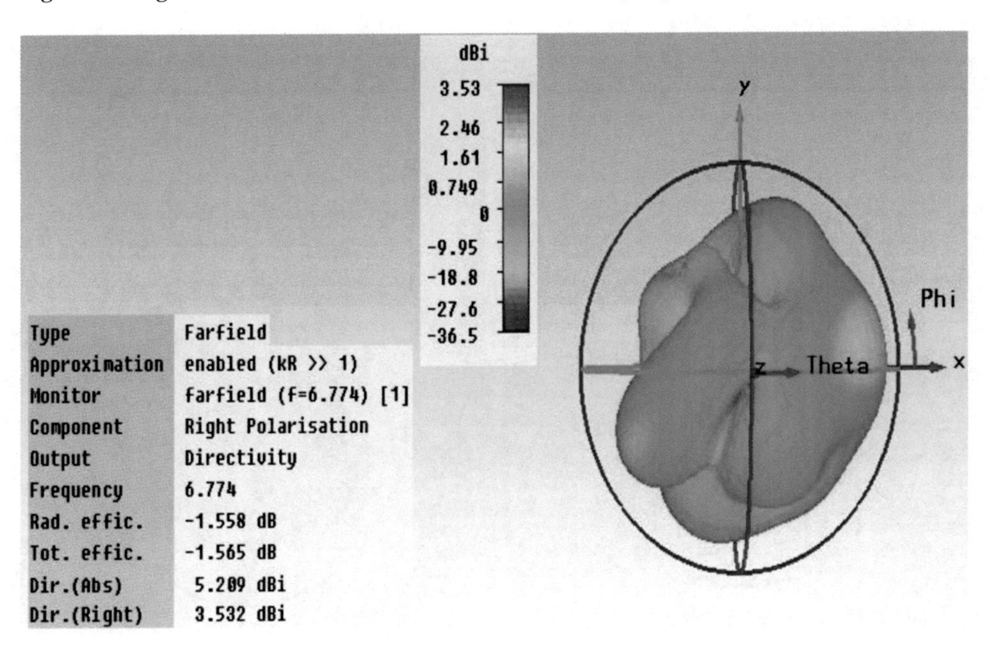

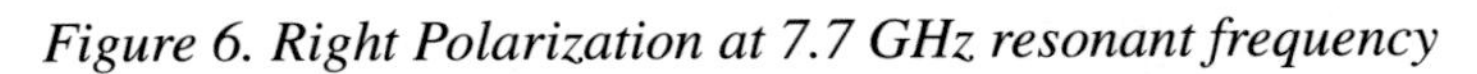

Figure 6. Right Polarization at 7.7 GHz resonant frequency

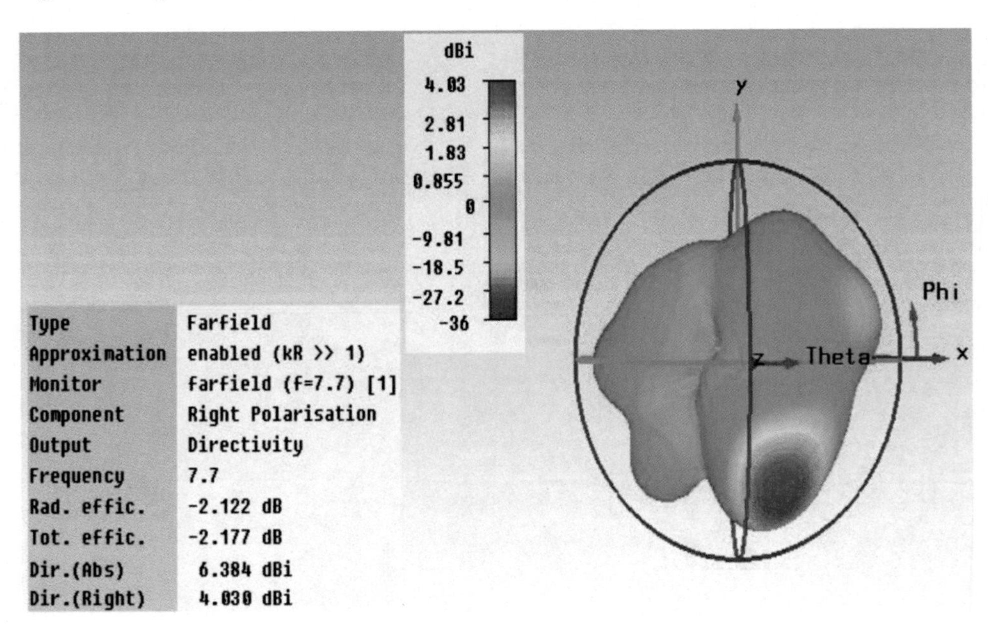

Figure 5. Left Polarization at 7.7 GHz resonant frequency

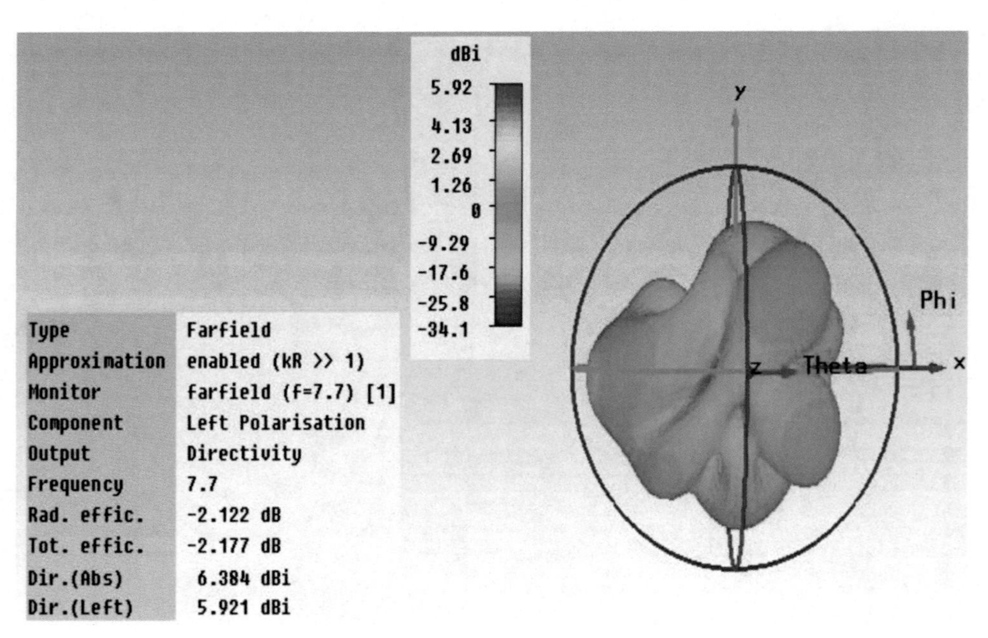

The directivity of left polarization at 6.77 GHz and 7.7 GHz frequency is 3.589 dBi and 5.921 dBi. The directivity of right polarization at 6.77 GHz and 7.7 GHz is 3.532 dBi and 4.030 dBi.

RESULTS AND DISCUSSION

After simulating the presented antenna various results are generated and shown in figure given below. Figure 7 shows the return loss Vs frequency graph in which 6.774 GHz and 7.7 GHz resonant frequency is generated and covers a band from 6.561 GHz to 8.07 GHz frequency. Figure 8 and Figure 9 shows the radiation pattern in 3-D format of presented antenna in which 5.209 directivity is generated on 6.774 GHz resonant frequency and 6.384 directivity is generated on 7.7 GHz resonant frequency. Also Figure 10 and Figure 11 show the radiation pattern in polar plot of presented antenna of two resonant frequency 6.774 GHz and 7.7 GHz. Also figure 12 and figure 13 shows the axial ratio of resonant frequencies 6.774 GHz and 7.7 GHz

Figure 7. Return loss Vs frequency simulated graph

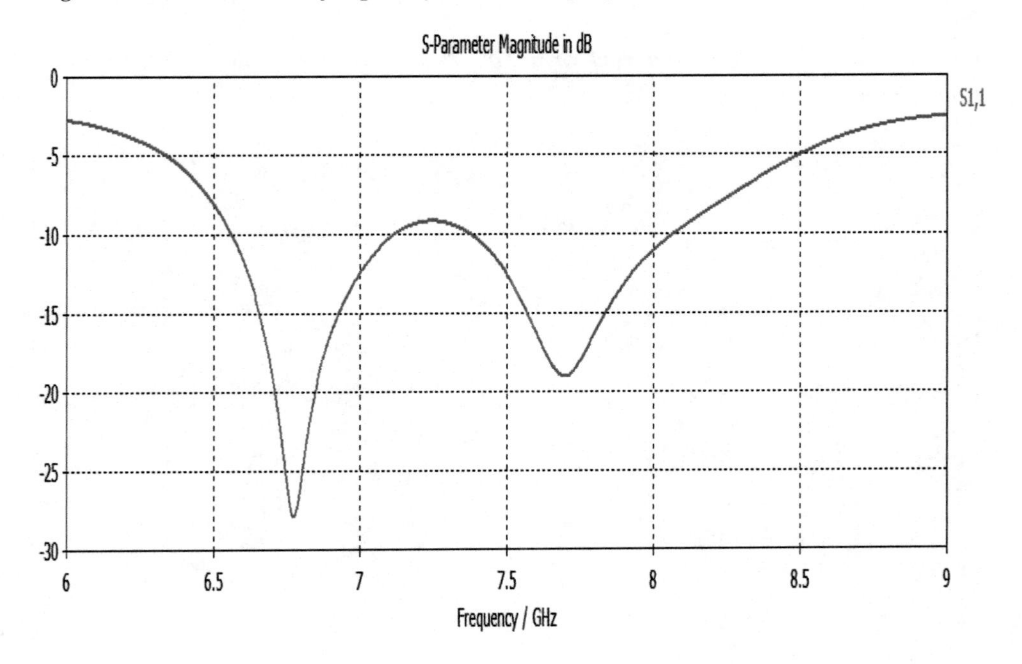

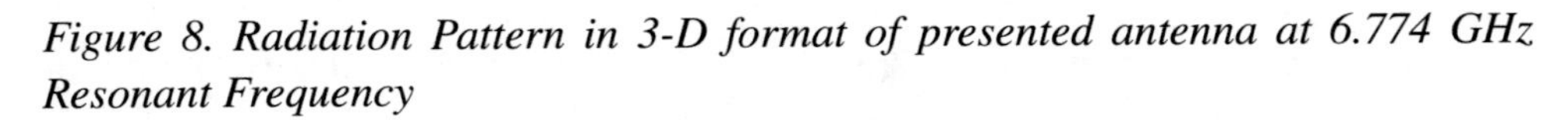

Figure 8. Radiation Pattern in 3-D format of presented antenna at 6.774 GHz Resonant Frequency

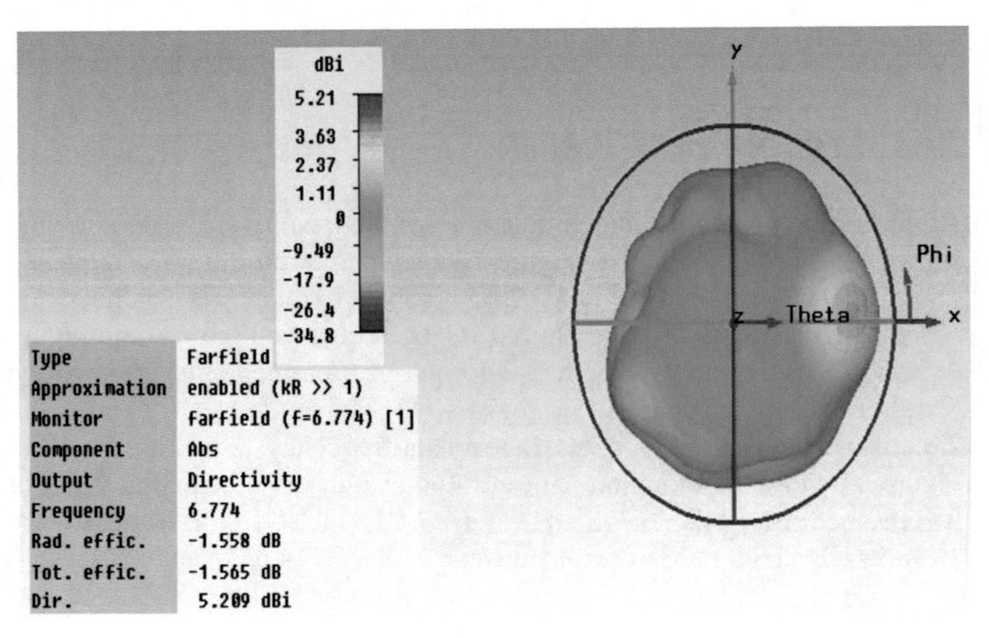

Figure 9. Radiation Pattern in 3-D format of presented antenna at 7.7 GHz Resonant Frequency

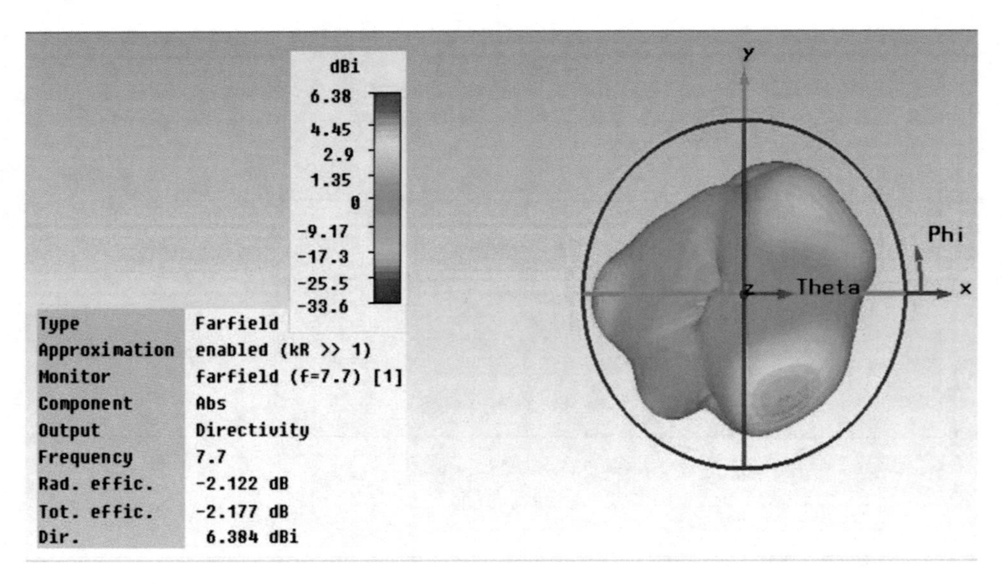

Figure 10. Radiation pattern in polar plot of presented antenna at 6.774 GHz Resonant Frequency

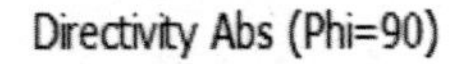

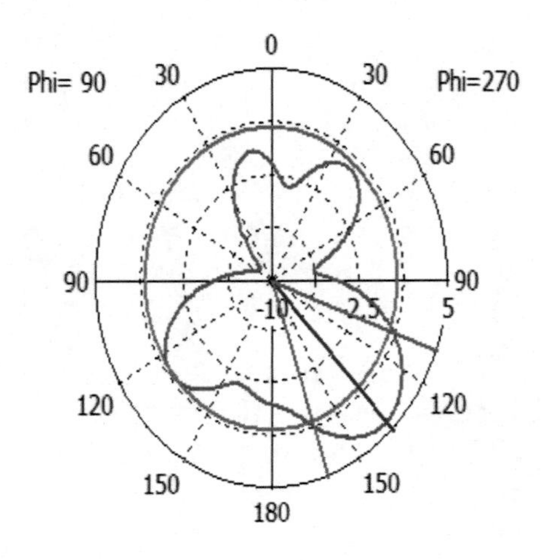

Figure 11. Radiation pattern in polar plot of presented antenna at 7.7 GHz Resonant Frequency

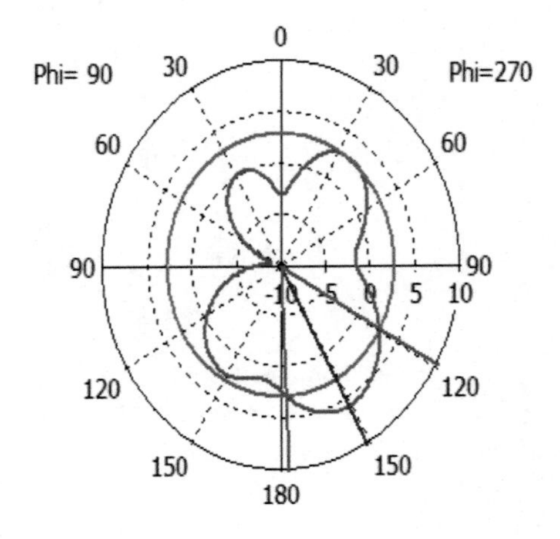

Figure 12. Axial Ratio at 6.774 GHz Resonant Frequency

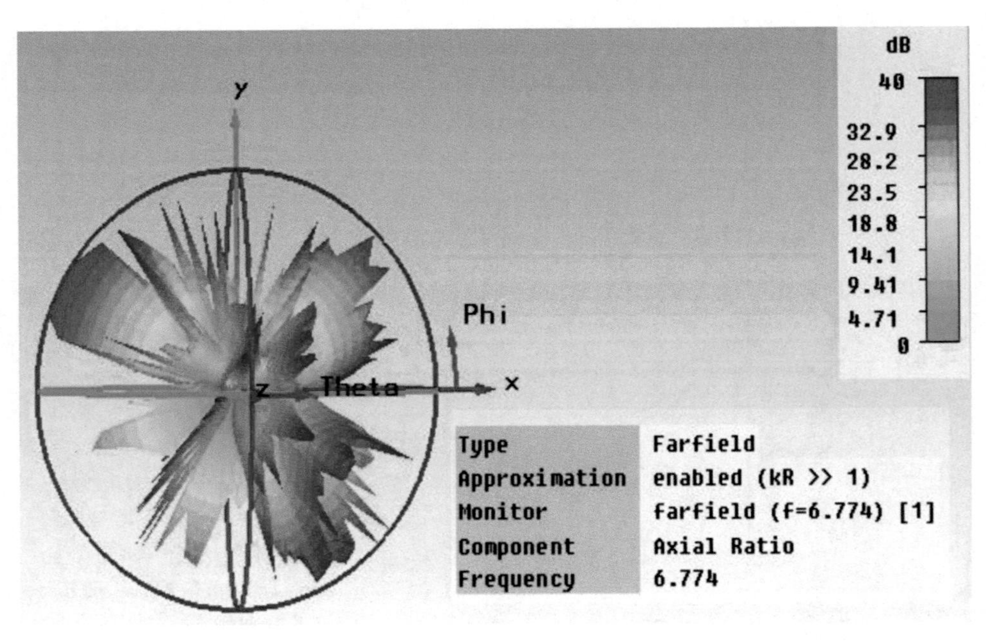

Figure 13. Axial Ratio at 7.7 GHz Resonant Frequency

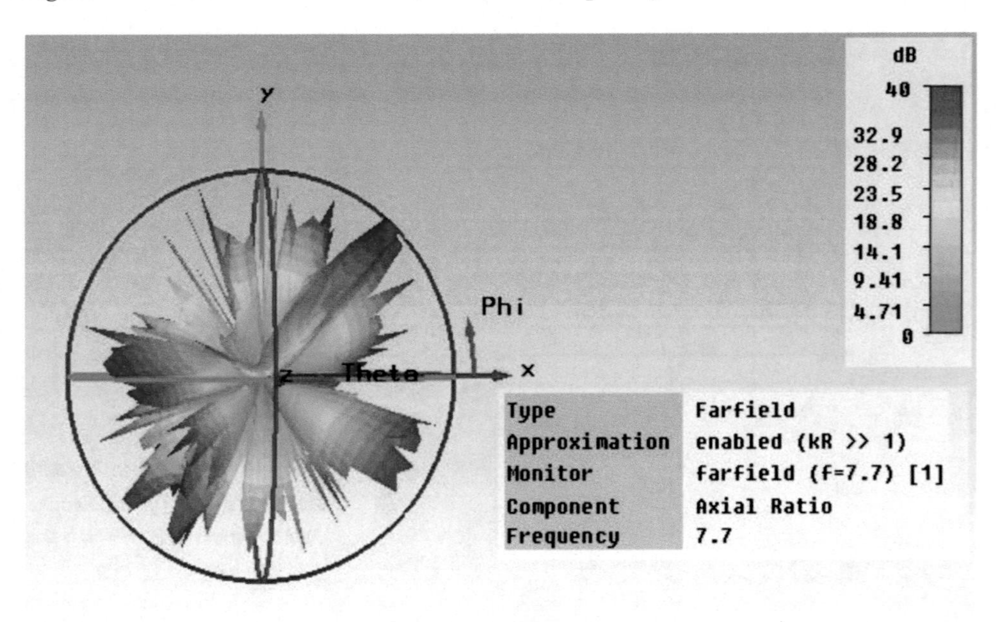

CONCLUSION

The proposed antenna was simulated and designed in CST software having two band from 6.555 Giga-Hertz to 7.122 Giga-Hertz and 7.38 Giga-Hertz to 8.07 Giga-Hertz. The anticipated antenna is useful for Multi Band Applications at resonant frequencies 6.774 Giga-Hertz and 7.7 Giga-Hertz. The overall presentation of proposed antenna is better than the presentation of other type of copper substrate antenna. Further this antenna is used for higher frequencies applications and can also be used for harvest the energy.

REFERENCES

Agilent HSMS -285x Series Surface Mount Zero Bias Schottky Detector Diodes-Data Sheet 5898-4022EN. (2005). Agilent Technologies, Inc.

Chahat, Zhadobov, & Sauleau. (2014). Antennas for Body Centric Wireless Communications at Millimeter Wave Frequencies. In *Progress in Compact Antennas*. Academic Press.

Chahat, Zhadobov, & Sauleau. (2014). Antennas for Body Centric Wireless Communications at Millimeter Wave Frequencies. In *Progress in Compact Antennas*. Academic Press.

Hai, S. Z., Shun, L. C., Xiao, K., & Liang, F. Y. (2013). Numerical and experimental analysis of Wideband e-shaped patch textile antenna. *Progress In Electromagnetics Research C*, *45*, 163–178. doi:10.2528/PIERC13091308

Hall, P. S., & Hao, Y. (2006). *Antennas and Propagation for Body-centric Wireless Communications*. Artech House.

Kennedy, Fink, Chu, Champagne, Lin, & Khayat. (2009). Body-worn e- textile antenna: The good, the low-mass, and the conformal. *IEEE Transaction on Antennas and Propagation, 57*(4), 910-918.

Lemey, S., Declercq, F., & Rogier, H. (2014). Textile Antennas as Hybrid Energy-Harvesting Platforms. *Proceedings of the IEEE*, *102*(11), 1833–1857. doi:10.1109/JPROC.2014.2355872

Lemey, S., Declercq, F., & Rogier, H. (2014). Textile Antennas as Hybrid Energy-Harvesting Platforms. *Proceedings of the IEEE*, *102*(11), 1833–1857. doi:10.1109/JPROC.2014.2355872

Lui, K. W., Murphy, O. H., & Toumazou, C. (2013). wearable wideband circularly polarized textile antenna for effective power transmission on a wirelessly-powered sensor platform. *IEEE Transactions on Antennas and Propagation*, *61*(7), 3873–3876. doi:10.1109/TAP.2013.2255094

Lui, K. W., Murphy, O. H., & Toumazou, C. (2013). Wearable wideband circularly polarized textile antenna for effective power transmission on a wirelessly-powered sensor platform. *IEEE Transactions on Antennas and Propagation*, *61*(7), 3873–3876. doi:10.1109/TAP.2013.2255094

Mane, Patil, & Dhanawade. (2014). Comparative Study of Microstrip Antenna for Different Subsrtate Material at Different Frequencies. *International Journal of Emerging Engineering Research and Technology, 2*(9).

Murali, Muni, Varma, & Chaitanya. (2014). Development of Wearable Antennas with Different Cotton Textiles. *Int. Journal of Engineering Research and Applications, 4*(7), 8-14.

Roundy, S. J. (2003). *Energy scavenging for wireless sensor nodes with a focus on vibration to electricity conversion* (PhD Thesis). University of California, Berkeley, CA.

Saxena, Singh, Mohini, Bhardwaj, Chae, Sharma, & Bhoi. (2018). Rectenna Circuit at 6.13 GHz to operate the sensors devices. *International Journal of Engineering and Technology, 7*(2.33), 644-646.

Saxena & Singh. (2018). A Watch E-Cut Shape Textile Antenna for WB Applications. *Journal of Microwave & Technologies*, *5*(1), 29–32.

Saxena & Singh. (2018). A Moon-Strip Line Antenna for Multi-Band Applications at 5.44 GHz Resonant Frequency. *4th International Conference on Advances in Electrical, Electronics, Information, Communication and Bio-Informatics (AEEICB-18).*

Shukla, S. S., Verma, R. K., & Gohir, G. S. (2015). Investigation of the effect of Substrate material on the performance of Microstrip antenna. *4th International Conference on Reliability, Infocom Technologies and Optimization*, 1-3. 10.1109/ICRITO.2015.7359350

Singh, Saxena, Khare, Shayka, Chae, Sharma, & Bhoi. (2018). Power Harvesting through flexible rectenna at dual resonant frequency for low power devices. *International Journal of Engineering and Technology, 7*(2.33), 647-649.

Singh, V. K., & Saxena, A. (2018). Two Parabolic Shape Microstrip Patch Antenna for Single Band Application. *Journal of Microwave & Technologies*, *5*(1), 33–36.

Van Torre, Vallozzi, Rogier, & Verhaevert. (2010). Diversity textile antenna systems for firefighters. *Antennas and Propagation (EuCAP), 2010 Proceedings of the Fourth European Conference on.*

Van Torre, Vallozzi, Rogier, & Verhaevert. (2010). Diversity textile antenna systems for firefighters. *Antennas and Propagation (EuCAP), 2010 Proceedings of the Fourth European Conference on.*

Verma, V., Singh, V. K., & Saxena, A. (2016). Optimization of Microstrip Antenna for WLAN and WiMax Lower Band Applications. *Journal of Microwave & Technologies*, *3*(2), 19–24.

Chapter 2
A Hammer Type Textile Antenna With Partial Circle Ground for Wide-Band Application

Anurag Saxena
S. R. Group of Institution Jhansi, India

Bharat Bhushan Khare
https://orcid.org/0000-0001-8755-9808
UIT RGPV Bhopal, India

ABSTRACT

In this chapter, a partial circle ground textile patch antenna for wideband applications with better bandwidth is presented. The simulated antenna is proposed on textile jeans substrate having dielectric constant of 1.7. The radius of textile jeans substrate antenna is 15 mm. The overall simulation of partial circle grounded shaped antenna has been done using CST simulation tool. The simulated antenna resonates at frequency 9.285 GHz with the reflection coefficient of -28 dB. It covers a bandwidth from 7.008 GHz to 9.64 GHz. It has maximum directivity of 4.540 dBi.

DOI: 10.4018/978-1-5225-9683-7.ch002

INTRODUCTION

Microstrip antenna comprises three most important parts which is substrate, patch, and ground. A dielectric substrate i.e. genes is sandwiched between radiating patch and partial ground plane. The conducting area is located on the dielectric substrate which is used as radiating element. On other side of dielectric substrate there is conducting layer used as ground part (Srivastava, Singh, Ali, and Singh, 2013; Wong, 2002; Singh and Naresh, 2015; Singh, Ali, Ayub, and Singh, 2014; Raghupatruni, Krishna, and Kumar, 2013; Balanis, 2004). There are a variety of methods for enhancing the bandwidth of textile microstrip antenna by expands the material depth, utilizing low dielectric material, using numerous feeding techniques and impedance matching. Both the bandwidth and the thickness of the antenna is contradictory assets i.e. enhancement in bandwidth increases the size & thickness of presented antenna (Srivastava, Singh and Avub, 2015; Loni, Avub and Singh, 2014; Singh, Ali, Avub, and Singh, 2014; Singg, Singh and Singh, 2014; Singh, Singh and Naresh, 2016; Gupta, Singh, Ali, Ahirwar, 2016; Din, Chakrabarty, Ismail, Devi and Chen, 2012).

Currently the fast improvement of modern communication systems is necessary for transportable devices for some important features which includes easy designing, light weight, small in size, compatible with microwave, millimeter wave integrated circuits, less production cost and easy fabrication of microstrip antennas. The microstrip antenna has abundant useful properties which includes tiny size, low-cost of the fabrication, light weight, ease of setting up but the main limitations of printed antennas remains their narrow bandwidth features which restrictions the range of frequency such that the antenna be able to work efficiently. In wireless communication system, microstrip antenna plays major role. These are used in high performance aircrafts, radar, missiles and other spacecraft. It has many advantages such as its light weight, simple structure, ease of addition and less cost. Microstrip antenna requires very less space for installation as these are simple and small in size (Baudh, Kumar, and Singh, 2013; Srivastava, Avub, Singh, 2014; Loni and Singh, 2015; Ddhupkariva and Singh, 2015; Singh, Singh and Naresh, 2015a; Singh, Singh and Naresh, 2015b; Ali, Singh, Kumar and Shahanaz, 2011).

ANTENNA DESIGN CONFIGURATION

Figure1 shows the layout of presented antenna which works for the resonant frequencies. Either using rectangular substrate, the substrate of simulated antenna is in the form of circle having radius 15 mm. If there is more than one resonant part available with each operating at its own resonant frequency then the overlapping of multiple resonance leads to broadband applications.

Figure 1. Configuration of the anticipated hammer-type circle ground antenna (a) front looks (b) back look

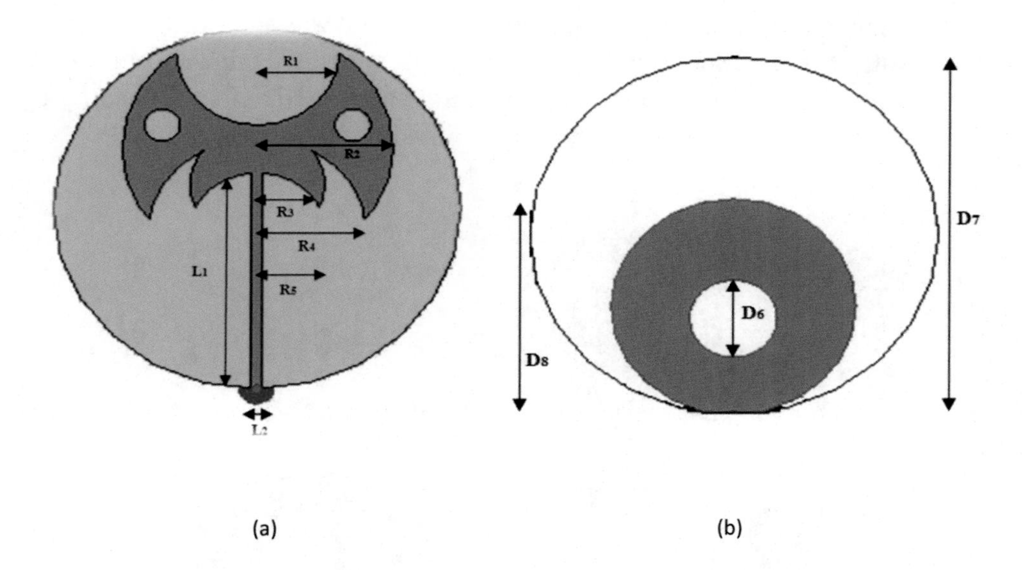

Table 1. Dimension of anticipated textile antenna

S.No	Parameter	Value
1	Substrate thickness	1mm
2	Relative permittivity	1.7
3	Circular substrate radius	15 mm
4	Upper circle cut radius (R_1)	6 mm
5	Outer circle patch radius (R_2)	10 mm
6	Lower circle cut radius (R_3)	5 mm
7	Inner two circle slot radius	1.3 mm
8	Strip line length (L_1)	12 mm
9	Strip line width (W_1)	0.8 mm

The basis of the proposed antenna was an annular ring which is cut from upper and lower side with outer patch radius (R_2) 10 mm and there are two circular slots which is at positive and negative half of x-axis having radius 1.3 mm. Table 1 shows the various parameters of anticipated antenna.

OPTIMIZATION OF PROPOSED MICROSTRIP ANTENNA

For the optimization and design of Partial Grounded Hammer Type antenna CST simulation software is used. In the beginning circular slot was used to achieve the desired bandwidth, then the circular slot has removed to modify the antenna patch and at last the projected shaped patch was achieved to optimize the results. Figure 2 shows the geometry of antenna1, antenna2 and antenna3. The return loss for antenna1, antenna2 and antenna3 has been denoted by blue, red and green colored curves respectively in the given figure 3.

Figure 2. Configurations of antennas (a) antenna1 (b) antenna2 (c) projected antenna3

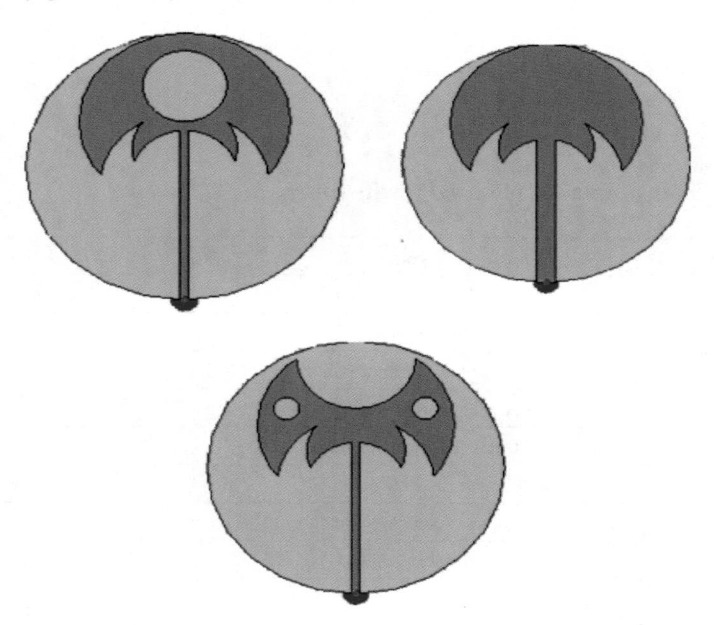

Figure 3. Simulated reflection coefficient vs frequency for the proposed textile microstrip antenna1, antenna2 & antenna3

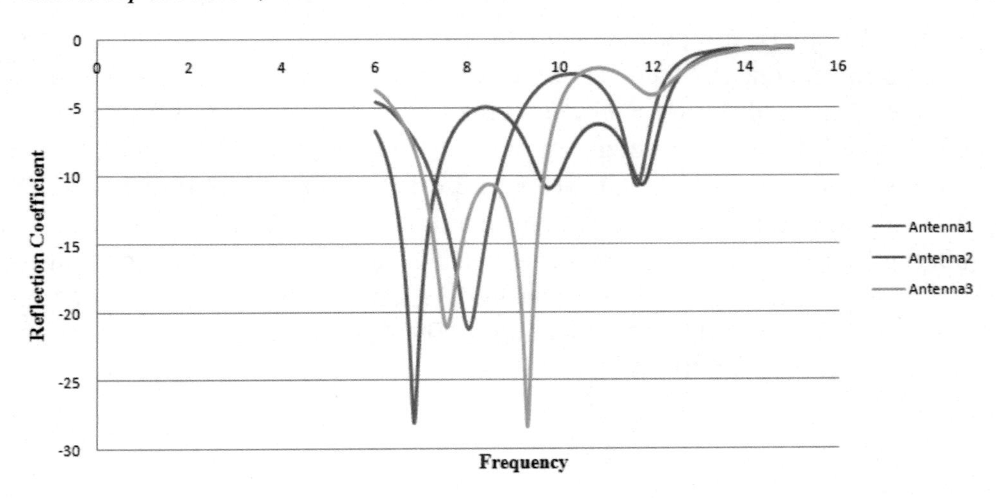

Figure 4. Computer-generated 2-D radiation pattern of anticipated textile microstrip antenna at 7.557 GHz

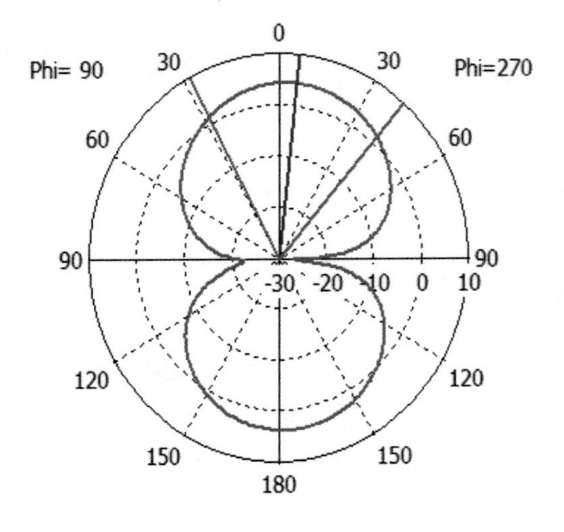

farfield (f=7.557) [1]

Frequency = 7.557
Main lobe magnitude = 4.4 dBi
Main lobe direction = 6.0 deg.
Angular width (3 dB) = 68.7 deg.

RESULTS AND DISCUSSION

Figure 4 depicts computer-generated 2-D radiation pattern of anticipated microstrip antenna at 7.557 GHz which describes the main lobe direction=6.0^0 and angular width (3dB) =68.7^0 and main lobe magnitude = 4.4 dBi at φ= 90. Figure 5 depicts computer-generated 2-D radiation pattern of anticipated microstrip antenna at 9.285 GHz which describes the main lobe direction=2.0^0 and angular width (3dB) =60.5^0 and main lobe magnitude = 4.6 dBi at φ= 90. Figure 1.6 shows computer-generated 3-D radiation pattern of anticipated microstrip antenna at 7.557 GHz & 9.285 GHz which gives good radiation efficiency of about 0.008484 dB and 0.009145 dB and also the directivity of 4.397 dBi and 4.540 dBi.

Figure 5. Computer-generated 2-D radiation pattern of anticipated textile microstrip antenna at 9.285 GHz

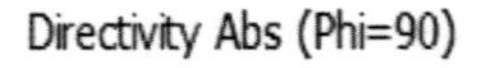

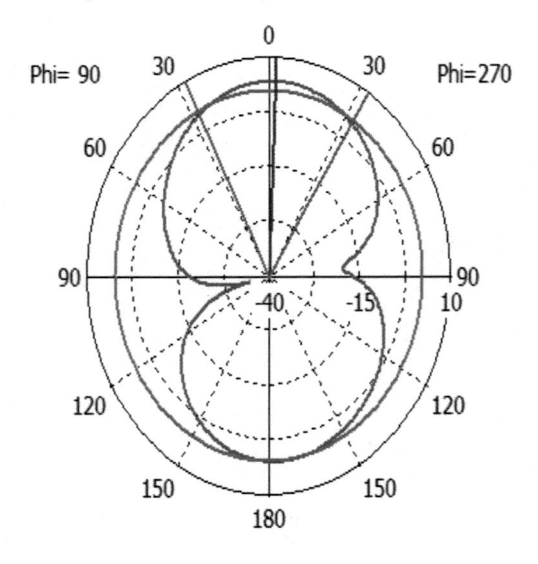

Figure 6. Simulated 3-D wave pattern of anticipated textile microstrip antenna at 7.557 GHz & 9.285 GHz

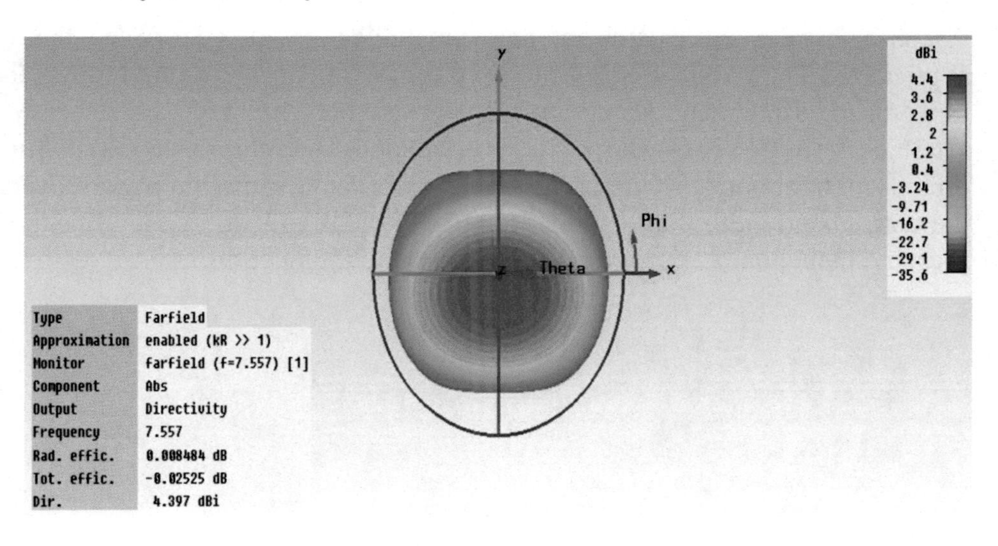

Figure 7.

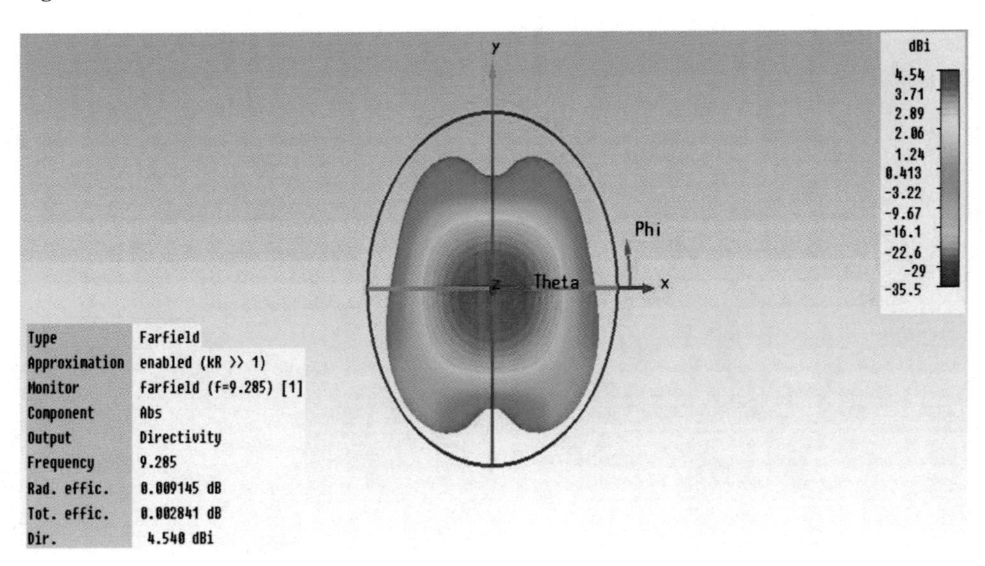

CONCLUSION

A jeans substrate textile patch antenna with circle is estimated for Wide band application. The radius of substrate is 15 mm with partial ground radius 10 mm presented antenna is very compact, thus suitable for installation on heavy satellites. Using CST software results are obtained. The antenna has maximum directivity of about 4.540 dBi. The antenna has stable radiation pattern over the entire bandwidth that extends from 7.557 GHz to 9.285 GHz, thus the presented antenna is appropriate for Wide Band application.

REFERENCES

Ali, Z., Singh, V. K., Kumar, A., & Shahanaz, A. (2011). E shaped Microstrip Antenna on Rogers Substrate for WLAN applications. *Proc. IEEE*, 342- 345. 10.1109/CICN.2011.72

Balanis, C. A. (2004). *Antenna Theory: Analysis and Design*. New York: John Wiley and Sons.

Baudh, R. K., Kumar, R., & Singh, V. K. (2013). Arrow Shape Microstrip Patch Antenna for WiMax Application. *Journal of Environmental Science, Computer Science and Engineering & Technology*, *3*(1), 269–274.

Dhupkariya, S., & Singh, V. K. (2015). Textile Antenna for C-Band Satellite Communication Application. *Journal of Telecommunication, Switching Systems and Networks*, *2*(2), 20–25.

Din, N. M., Chakrabarty, C. K., Bin Ismail, A., Devi, K. K. A., & Chen, W. Y. (2012). Design of RF Energy Harvesting System For Energizing Low Power Devices. *Progress in Electromagnetics Research*, *132*, 49–69. doi:10.2528/PIER12072002

Gupta, N., Singh, V. K., Ali, Z., & Ahirwar, J. (2016). Stacked Textile Antenna for Multi Band Application Using Foam Substrate. *Procedia Computer Science*, *85*, 871–877. doi:10.1016/j.procs.2016.05.277

Loni, J., Ayub, S., & Singh, V. K. (2014). Performance analysis of Microstrip Patch Antenna by varying slot size for UMTS application. *IEEE Conference on Communication Systems and Network Technologies (CSNT-2014)*, 1-5. 10.1109/CSNT.2014.9

Loni & Singh. (2015). Development of Bandwidth Enhanced Microstrip Patch Antenna for UMTS Application. *Journal of Microwave Engineering & Technologies, 2*(1), 1-7.

Raghupatruni, V. (2013). Design of Temple Shape Slot Antenna for Ultra Wideband Applications. *Progress In Electromagnetic Research B*, *47*, 405–421. doi:10.2528/PIERB12111001

Rawat, A. K., Singh, V. K., & Ayub, S. (2012). Compact Wide band Microstrip Antenna for GPS/WLAN/WiMax Applications. *International Journal of Emerging Trends in Engineering and Development*, *7*(2), 140–145.

Singh, Singh, & Naresh. (2015). Rectangular Slot Loaded Circular Patch Antenna for WLAN Application. *Journal of Telecommunication, Switching Systems and Networks, 2*(1), 7-10.

Singh, N., Singh, A. K., & Singh, V. K. (2015). Design & Performance of Wearable Ultra Wide Band Textile Antenna for Medical Applications. Microwave and Optical Technology Letters, 57(7), 1553-1557.

Singh, N. K., Singh, V. K., & Naresh B. (2016). Textile Antenna for Microwave Wireless Power Transmission. *Procedia Computer Science*, *85*, 856–861. doi:10.1016/j.procs.2016.05.275

Singh, V. K., Ali, Z., Ayub, S., & Singh, A. K. (2014). *A wide band Compact Microstrip Antenna for GPS/DCS/PCS/WLAN Applications. In Intelligent Computing, Networking, and Informatics* (Vol. 243, pp. 1107–1113). Springer.

Singh, V. K., Ali, Z., Ayub, S., & Singh, A. K. (2014). Bandwidth Optimization of Compact Microstrip Antenna for PCS/DCS/Bluetooth Application. *Central European Journal of Engineering, Springer*, *4*(3), 281–286.

Singh & Naresh. (2015). Multi Resonant Microstrip Antenna with Partial Ground for Radar Application. *Journal of Telecommunication, Switching Systems and Networks, 2*(1), 1-5.

Srivastava, R., Ayub, S., & Singh, V. K. (2014). Dual Band Rectangular and Circular Slot Loaded Microstrip Antenna for WLAN/GPS/WiMax Applications. *IEEE Conference on Communication Systems and Network Technologies (CSNT-2014)*, 45 - 48.

Srivastava, R., Singh, V. K., & Ayub, S. (2015). *Comparative Analysis and Bandwidth Enhancement with Direct Coupled C Slotted Microstrip Antenna for Dual Wide Band Applications. In Frontiers of Intelligent Computing: Theory and Applications* (Vol. 328, pp. 449–455). Springer.

Srivastava, S., Singh, V. K., Ali, Z., & Singh, A. K. (2013). Duo Triangle Shaped Microstrip Patch Antenna Analysis for WiMAX lower band Application. *Procedia Technology, 10*, 554-563. 10.1016/j.protcy.2013.12.395

Wong. (2002). *Compact and Broadband Microstrip Antennas*. John Wiley & Sons, Inc.

Chapter 3
Design of RF Rectenna on Thin Film to Power Wearable Electronics

Naresh B.
Bhagwant University, India

Vinod Kumar Singh
https://orcid.org/0000-0002-0671-0631
SRGI Jhansi, India

Virendra Kumar Sharma
Bhagwant University, India

ABSTRACT

This chapter reports a hybrid wearable energy harvesting system. Integration of microwave antenna on thin film amorphous silicon solar cell creates a hybrid system that can harvest both the solar and microwave energies. The antenna designed on solar cell will harvest the microwave energy at dual frequencies 1.85 GHz and 2.45 GHz with an effective return loss of 28dB and 27dB respectively. A complete hybrid harvesting system consist of a flexible solar cell, antenna, voltage doubler, and impedance matching dual band filter. The rectifier, designed on a rigid glass-reinforced epoxy substrate, is a voltage doubler and a matching circuit is designed by microstrip lines is used.

DOI: 10.4018/978-1-5225-9683-7.ch003

INTRODUCTION

Power consumption of Semiconductor devices are significantly reduce and now these device are operate at ultra-low power levels. So, that energy harvested from environment will sufficient to run the electronic circuits. Much of electronic devices are compact in size, handy and coming with more powerful features which made daily life comfortable. Nowadays wearable technology is developing area based on energy harvested being used to power the integrated devices into clothing like sensors, LED displays and to charge the super capacitors etc.

Mostly the energy can be harvested from solar, piezoelectric, microwave (Radio Frequency) energy sources to power low, ultra-low power electronics, sensors or to recharge thin super capacitors. The energy harvesting technology will reduce weight burden on dismounted soldiers in terms of batteries to carry, no maintenance cost, all mostly power can regenerate from environment which extend the charging cycle. Researchers are inventing new technologies, materials which has less weight and flexible to wear or integrate into clothing. Flexible solar cell based energy harvesting have been reported using pollymide thin film photovoltaic smart bracelet for healthcare application, measuring the heartbeats of a patient through photoplethysmography (PPG) (Jokic and Magno, 2017; Wu, Arefin, Redoute and Yuce, 2017). Inductively power transfer circuit is designed on plastic with flexible solar cell as a power source (Hu et al., 2012), an intelligent hardware based charging controller circuit is presented for wireless sensor application (Li, Yin, Au-Shi, and Ronghua, 2015).

The conversion of ambient microwave power into DC power has been done by a circuit called rectenna. Usually rectenna consists of a radio frequency (RF) receiving antenna and a rectifying circuit. Solid dielectric substrate antennas are more common in use, they made with a printed copper on a dielectric substrate. The concept of wearable or textile antenna are having possible way to power the incorporated electronic devices with maximum efficiency.

In hybrid energy harvesting system energy can be harvesting from multiple sources, such as a low cost compact solar/electromagnetic harvester designed on flexible polyester (Vera, Gianfranco, Georgiadis, Apostolos, Collado, Ana, and Via, 2010), dual band rectenna was implemented on flexible polyethylene terephthalate PET substrate (Collado and Georgiadies, 2013), multi-input multi output three source solar, vibration and thermal harvester in (Bandypadhyay and Chandrakasan, 2012), vibration based micro- electrometrical energy harvester (Aljibori, Hakim, Salim, Dhia, Haris, and Kherbeet, 2015).

In this paper a wearable and flexible photovoltaic dual band RF energy harvester is proposed to power up the wearable electronic gadgets. Dual band rectenna can harvest the microwave energy at 1.85 GHz and 2.45 GHz bands with power level starts from -20 dBm. A thin film amorphous solar cell encapsulation was used as substrate for the dual band hollow rectangle antenna, without effecting the solar cell performance.

FLEXIBLE SOLAR MICROWAVE ANTENNA

Thin Film Solar Cell

Flexible thin film solar system even through introduced in year 1970's due to low efficiency they are not much popular that of crystalline silicon solar system. Nowadays the efficiency of the thin film solar cell are almost same as the rigid silicon panels and their flexible characteristic, light weight, easier installation are main motives for flexible solar system (Chopra and Das, 1983; Chapin, Fuller and Pearson, 1954; Chopra and Das, 1983b; Birkmire and Eser, 1997; Fahrenbruch and Bube, 1983). In this thin film solar technology amorphous silicon (a-Si:H) module is receiving most attention due to absorb sunlight very efficiently, the overall thickness of the layer is less than 1μm. Therefore the substrate which supported these layers also much thicker which is way overall module is thin and flexible.

Amorphous silicon (a-Si:H) is the non-crystalline form of the semiconductor silicon. The fabrication process of the thin film solar cell is; silicon material

Table 1. Solar cell description

S.No	Parameter	Value
1	Maximum power	0.5W
2	Operating voltage	1.5V
3	Operating current	0-330 mA
4	Short circuit current	420 mA
5	Open circuit voltage	2.0 V
6	Solar cell Size	198 X 58 X 0.8 mm

vapor made to deposit upon a 1 micrometer thick surface such as metal, glass and plastic. The solar cell used in this paper is three junction amorphous thin film, the multilayer junction will increase the sun light absorption from the full solar spectrum. The solar cell electrical parameters and dimension are presented in Table1.

Design of Antenna on Solar Cell

The planar microstrip antenna is a low weight, easy to design and fabrication, wideband and multiband operation are obtained at low cost. The way microstrip antennas being incorporated with modern electronics for communication; the same we can power the wearable electronics. There are two ways of design methods in RF solar energy harvesting, first one is autonomous integration of antenna with solar cell, in which solar cell is a separate power source one require additional surface area to implant. The second one is full integration of antenna with solar cell, in which antenna is designed upon the solar cell without effecting the solar efficiency.

In this paper, to power the wearable electronics full integration technique was used; which means thin film solar cell encapsulation layers which involves ethylene-vinyl acetate (EVA) and nylon fabric are used as substrate for the microwave antenna. The thin film solar cell is placed between stacked layers of protective transparent layer on top and a dielectric layers at the bottom,

Figure 1. Five layer encapsulation

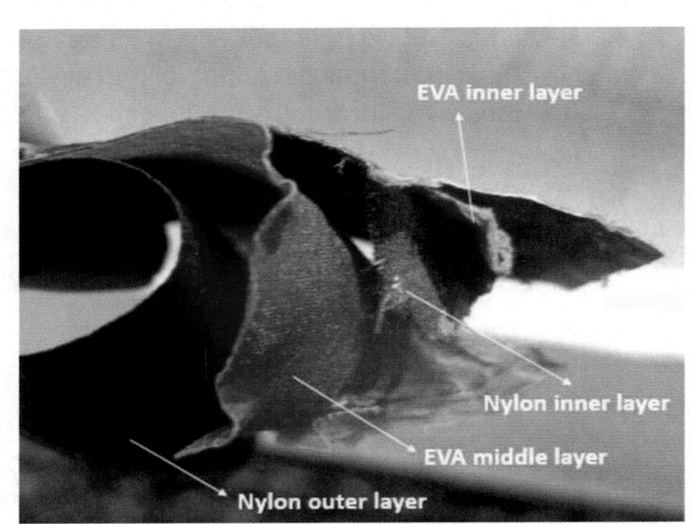

the aim of the encapsulation is protecting the solar cell from environmental degradation and ensures its operation and reliability. The encapsulation layers individual are shown in Figure 1.

The microstrip antenna designed on solar cell has a rectangle ring shape. Without effecting the electrical efficiency of solar cell radiating element lays on boundaries of the solar cell and in this way multi source harvesting is possible. Microstrip antenna required mainly required radiating element and dielectric substrate, encapsulation of the solar cell have the dielectric material and the radiating rectangular ring is designed by a copper foil tape.

The microstrip antenna designed on solar cell has a rectangle ring shape. Without effecting the electrical efficiency of solar cell radiating element lays on boundaries of the solar cell and in this way multi source harvesting is possible. Microstrip antenna required mainly required radiating element and dielectric substrate, encapsulation of the solar cell have the dielectric material and the radiating rectangular ring is designed by a copper foil tape. The antenna designed in CST software is shown in figure.2; the physical dimensions of the antenna are presented in Table 2.

Table 2. Antenna parameters

S.No	Parameter	Value
1	W [mm]	196
2	W1 [mm]	194
3	L [mm]	58
4	L1 [mm]	50
5	Feed line [mm]	8X2
6	Ground [mm]	24X189

Figure 2. CST antenna model

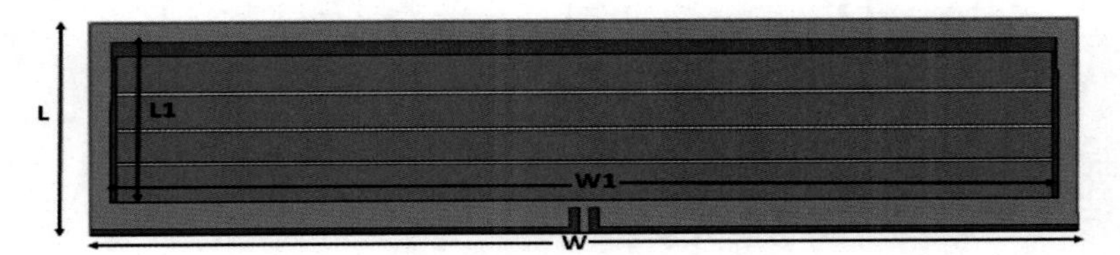

VOLTAGE DOUBLER AND IMPEDANCE MATCHING

There are several types of topologies in rectification of RF to DC such as single series, single shunt, full wave diode rectifiers (Marian, Allard, Vollaire, and Verdier, 2012) and voltage double rectifiers. The efficiency of the rectenna depends on frequency, RF diode type, input power level, matching between antenna and rectifier circuit; power losses if rectifier offers minimum losses electrical efficiency is high. Therefore standard voltage doubler is the noble option over half wave rectifier at ultra-low power levels. Microwave signal arrived at the input of the rectifier $v_{in} = V_{AC} \sin \omega t$ where V_{AC} is the amplitude and ω is the frequency of the input signal.

The fig.3 explains the rectification of microwave positive and negative half cycle at voltage doubler. The diode D_1 comes into forward bias to negative half wave and capacitor C_1 charges, the diode D_2 rectifies the positive half wave and capacitor C2 is charges. Finding the input impedance of the RF (voltage doubler) diode for design of matching circuit is the vital point since diode has non liner characteristics. The output DC voltage (V_{DC}) at the load is given by eq(1) and input impedance of the voltage doubler is calculate by using eq(2) (Song, Chaoyun, Hung, Yi, Carter, Paul, et al., 2016), where I_S is the diode saturation current, m is the ideality factor and V_T is the thermal voltage.

$$V_{DC} = 2V_{AC} \sin \omega t - 2V_F \tag{1}$$

Figure 3. Voltage doubler

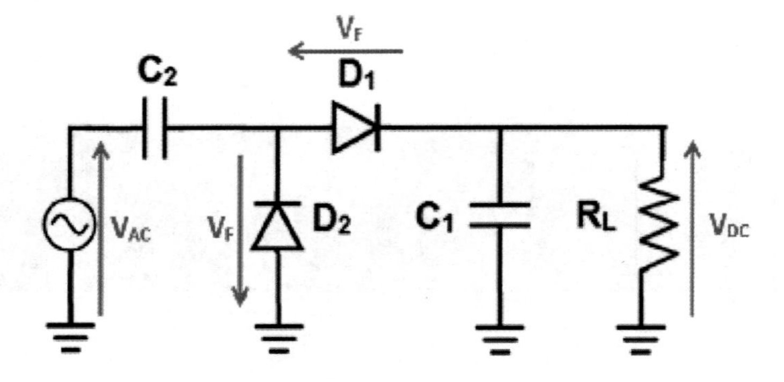

$$Z_D = \frac{V_{AC} \sin \omega t - 0.5 V_{DC}}{I_S \left[B_0 \left(\frac{V_{AC}}{m V_T} \right) \exp \left(\frac{-0.5 V_{DC}}{m V_T} \right) - 1 \right]} \quad (2)$$

The voltage doubler circuit was designed in ADS software is shown in Figure 4, analyzed with large signal S-parameter simulation to verify the power level dependent and non-liner characteristics behavior of the Schottky diode SMS7630 (Surface, 2013). Input impedance of the voltage doubler (Z_{VD}) obtained from ADS simulation is depicted in Figure 5, at 1.85 GHz the impedance is Z_{VD} = 43.071-j301.519 and 2.45 GHz it is Z_{VD} = 28.963-j228.833 with -30dBm as input power for a load resistance of 25K ohms. It is clear that real part of the voltage doubler is effected by the frequency, it changes from 43.071Ω to 28.963Ω; the imaginary part effected more it rapidly changes from 301.519Ω to 228.833Ω.

The primary goal of the matching circuit is minimize the reflection from the voltage doubler, maximize the received microwave power and matching is necessary for power transfer from one stage to another (Pavone, Buonanno, D'Urso, and Corte, 2012; Chaudhary, Kim, Jeong and Yoon, 2012; Hameed, Zohaib, and Moez, 2017; Wang, xinhuai and Zhang, 2013).

In the microwave frequency region, filters can be designed using distributed transmission lines. Series inductors and shunt capacitors can be realized with microstrip transmission lines. ADS Smith chat utility is used to design the filters by considering the source impedance as 50 Ω and load impedance is the diode impedance at respective frequencies.

Figure 4. Voltage doubler block designed in ADS software

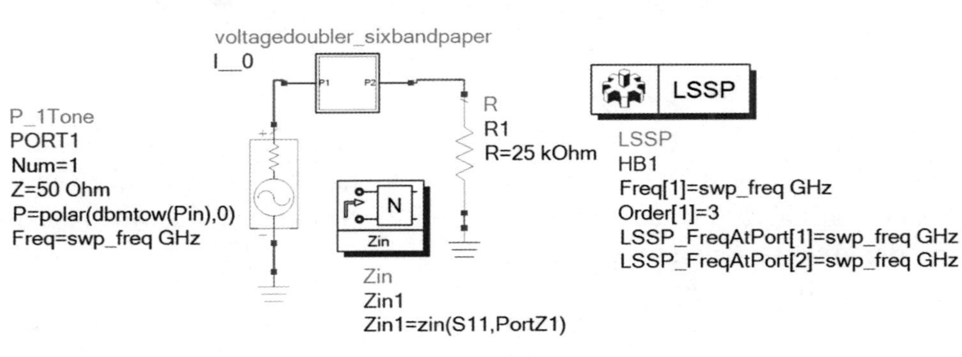

(a)

Figure 5. Simulated input impedance of the voltage doubler

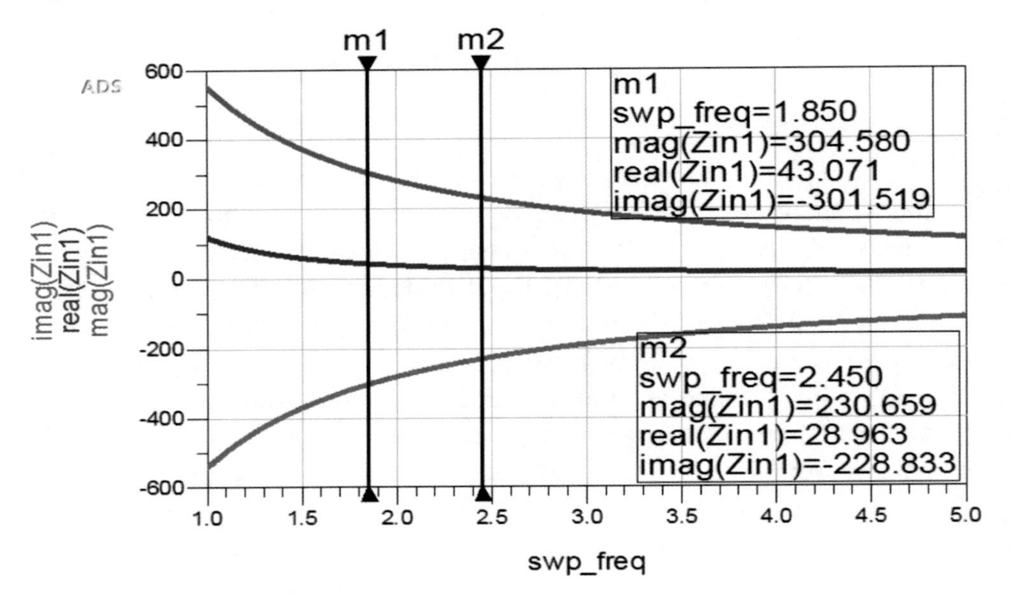

Figure 6. Simulated S11 plot of matching circuit at 2.45GHz and 1.86 GHz

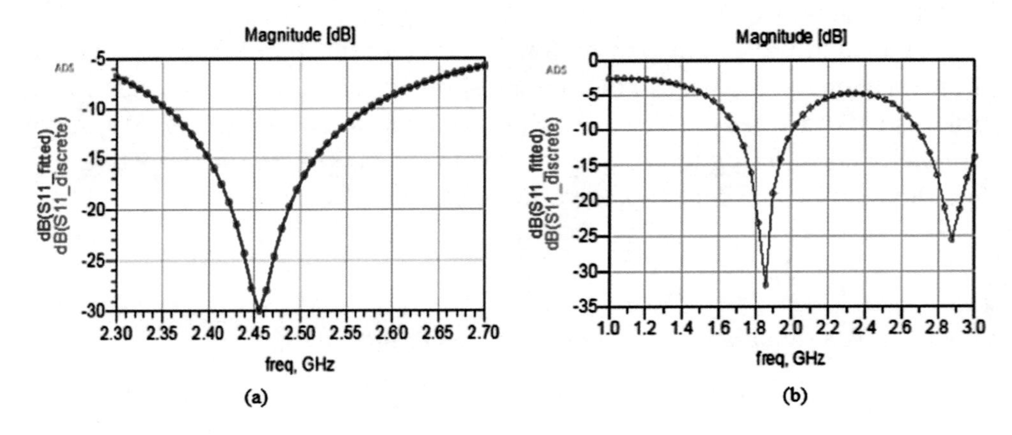

The simulated return loss plot of the matching circuits was shown in Figure 6(a) and 6(b). Both the matching circuit has the reflection coefficient (S11) -30dB at resonance frequencies, the S21 is near the 0 dB which means the signal noise generated by the nearby components will be rejected hence losses will be reduced.

Large-signal S-parameter (LSSP) is a non-linear simulation that accounts for power level-dependent behavior of the diode. The LSSP simulation of

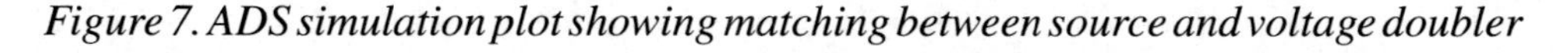

Figure 7. ADS simulation plot showing matching between source and voltage doubler

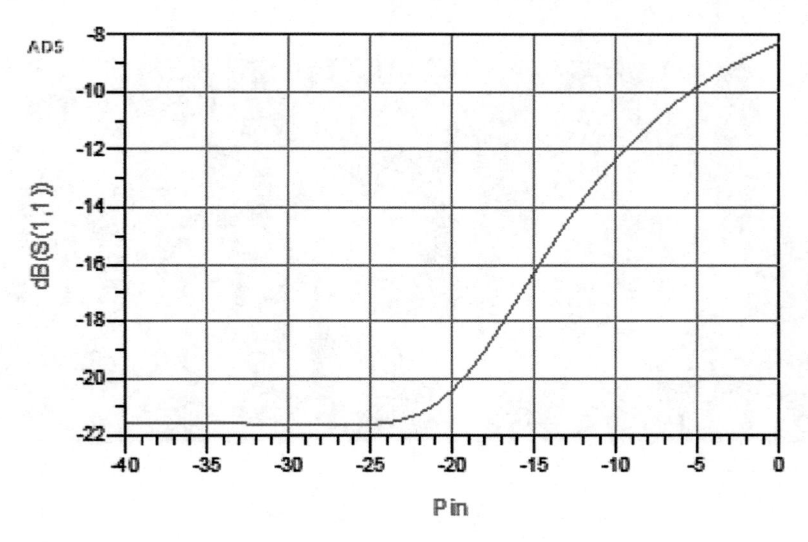

the impedance circuits is carried out to measure degree of matching between source and voltage doubler against microwave input power level. Unlike small-signal S-parameters, which are based on a small-signal simulation of a linearized circuit, large-signal S-parameters are based on a harmonic balance simulation of the full nonlinear circuit. Because harmonic balance is a large-signal simulation technique, its solution includes nonlinear effects. Figure. shouldee7 shows the result obtained from simulation, a very good range of matching was obtained, from -30 dBm to -15 dBm the S11 magnitude is below -15dB.

EXPERIMENTAL SETUP AND RESULTS

The antenna measurement is done to find the parameters of the antenna in both bent and unbent conditions using keysight microwave analyzer. The snapshot of the experimental setup is shown in Figure.8 (a). The measured return loss with the help of microwave analyzer is shown in Figure.8(b). The measured result shows that the antenna has dual band nature; The energy harvesting frequencies are 2.45 GHz and 1.86 GHz bandwidths are 1.70 – 1.90 GHz (200MHz) and 2.35- 2.48 GHz (130Mhz). The primary band center resonance frequency is 1.85 GHz with return loss magnitude of -27dB. The

Figure 8. (a) Return loss testing of designed flexible antenna (b) screen shot of measured S11 in dB.

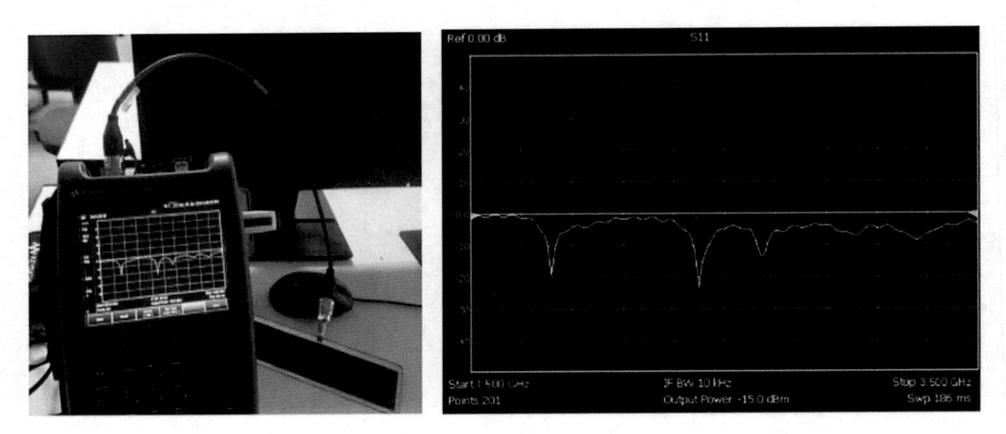

Figure 9. Wearable rectenna on thin film solar cell

resonance frequency of the second band is 2.45 GHz with the return loss magnitude of -28 dB.

In this study, voltage doubler long with matching circuit was built on PCB with 1.6 mm thickness and its conversion efficiency performance at resonance frequencies was studied when the input RF power was varied, the hardware model is shown in fig.9. Two zero-bias Schottky diode (SMS7630) were chosen in this design due to its low forward bias voltage and high saturation current.

A non-linear spice model with parasitic elements, provided by Skyworks Solution. Further, the bypass capacitors (C1, C2) were chosen to be 100 pF and storage capacitors (C3) 100 μF, both from Panasonic Electronics. The conversion efficiency, η, of the voltage doubler for RF power levels varying

Figure 10. (a) Simulated and measured rectifier RF-DC conversion efficiency with load resistance 7K ohms.(b) Simulated and measured rectifier RF-DC conversion efficiency for different load resistances

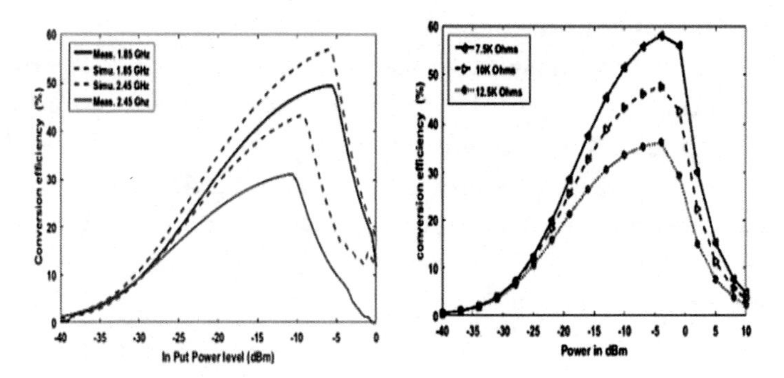

Figure 11. conversion efficiency in terms of load resistance

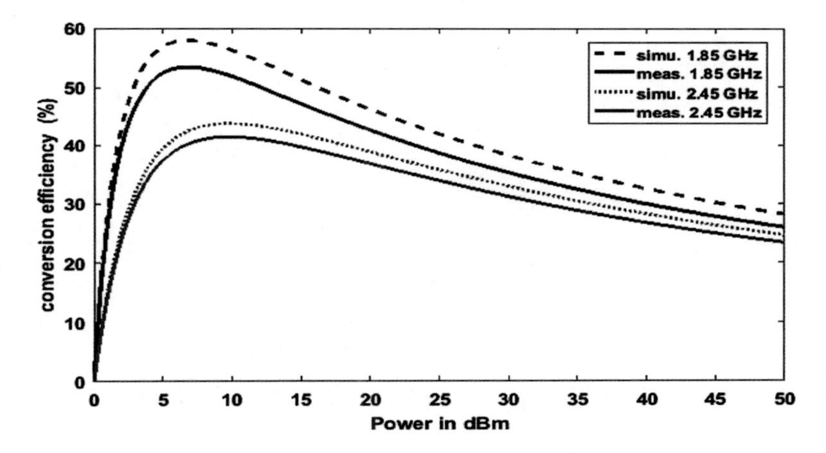

from -40 to 0 dBm is shown in Figure.7 at harvesting frequencies. The voltage doubler conversion efficiency at 1.85 GHz and 2.45 GHz are 58% and 43% respectively for a load of 7K Ohms shown in figure 10(a). In figure.10 (b) the conversion efficiency was plotted against power for different load resistance and maximum efficiency was obtained for 7K Ohms.

The load-dependent conversion efficiency at two frequencies for input power level -5dBm is depicted in Figure.11. It can be seen that the efficiency is greater than 50% at 1.85 GHz and 40% at 2.45 GHz, respectively for the load resistance between 5 kΩ and 10 kΩ.

CONCLUSION

This paper work explains the design and testing of an antenna with thin film solar cell encapsulation materials and also its potential application in the wearable wireless energy harvesting system. The solar cell encapsulation materials are EVA, nylon and top layer is ETFE; thus copper tap as a radiating element. From the Measured results the fabricated antenna has a dual band operation and the operation frequencies are 1.85 GHz and 2.45 GHz. Proposed solar cell rectenna is tested for different micropower levels as -40dBm to 10dBm. The maximum conversion efficiency obtained is 58% at -5dBm input power with DC output voltage of 400mV.

REFERENCES

Bandyopadhyay, S., & Chandrakasan, A. P. (2012). Platform Architecture for Solar, Thermal, and Vibration Energy Combining With MPPT and Single Inductor. *IEEE Journal of Solid-State Circuits*, *47*(9), 2199–2215. doi:10.1109/JSSC.2012.2197239

Birkmire, R. W., & Eser, E. (1997). Polycrystalline thin-film solar cells: Present status and future potential. *Annual Review of Materials Science*, *27*(1), 625–653. doi:10.1146/annurev.matsci.27.1.625

Chapin, D. M., Fuller, C. S., & Pearson, G. L. (1954). A new silicon p–n junction photo cell for converting solar radiation into electricalpower. *Journal of Applied Physics*, *25*(5), 676–677. doi:10.1063/1.1721711

Chaudhary, G., Kim, P., Jeong, Y., & Yoon, J. H. (2012, October). Design of high efficiency RF-DC conversion circuit using novel termination networks for RF energy harvesting system. *Microwave and Optical Technology Letters*, *54*(10), 2330–2335. doi:10.1002/mop.27087

Chen & Jin. (n.d.). Piezoelectric and electromagnetic hybrid energy harvester for powering wireless sensor nodes in smart grid. *Journal of Mechanical Science and Technology.*

Chopra, Paulson, & Dutta. (2004). Thin-Film Solar Cells: An Overview. *Progress in Photovoltaics, 12*, 69-92. . doi:10.1002/pip.541

Chopra, K. L., & Das, S. R. (1983). *Thin-Film Solar Cells*. New York: Plenum. doi:10.1007/978-1-4899-0418-8

Collado, A., & Georgiadis, A. (2013, August). Conformal Hybrid Solar and Electromagnetic (EM) Energy Harvesting Rectenna. *IEEE Transactions on Circuits and Systems. I, Regular Papers*, *60*(8), 2225–2234. doi:10.1109/TCSI.2013.2239154

Fahrenbruch, A. L., & Bube, R. H. (1983). *Fundamentals of Solar Cells*. New York: Academic Press.

Gianfranco, Georgiadis, Collado, & Via. (2010). Design of a 2.45 GHz rectenna for electromagnetic (EM) energy scavenging. *IEEE Radio and Wireless Symposium, RWW 2010*, 61 - 64. 10.1109/RWS.2010.5434266

Hameed, Z., & Moez, K. (2017). Design of impedance matching circuits for RF energy harvesting systems. *Microelectronics Journal*, *62*, 49–56. doi:10.1016/j.mejo.2017.02.004

Hu, Y. (2012). Flexible solar-energy harvesting system on plastic with thin-film LC oscillators operating above ft for inductively-coupled power delivery. *Proceedings of the IEEE 2012 Custom Integrated Circuits Conference*, 1-4.

Jokic, P., & Magno, M. (2017). Powering smart wearable systems with flexible solar energy harvesting. *2017 IEEE International Symposium on Circuits and Systems (ISCAS)*. 10.1109/ISCAS.2017.8050615

Li, Yin, Au-Shi, & Ronghua. (2015). An intelligent solar energy-harvesting system for wireless sensor networks. *EURASIP Journal on Wireless Communications and Networking.*

Marian, V., Allard, B., Vollaire, C., & Verdier, J. (2012). Strategy for Microwave Energy Harvesting From Ambient Field or a Feeding Source. *IEEE Transactions on Power Electronics*, *27*(11), 4481–4491. doi:10.1109/TPEL.2012.2185249

Pavone, D., Buonanno, A., D'Urso, M., & Corte, F. (2012). Design considerations for radio frequency energy harvesting devices. *Progress In Electromagnetics Research B*, *31*, 19–35. doi:10.2528/PIERB12062901

Salim, Salim, & Khir, Haris, & Kherbeet. (2015). A review of vibration-based MEMS hybrid energy harvesters. *Journal of Mechanical Science and Technology*, *11*(1), VL–29.

Song, C., Huang, Y., Carter, P., Zhou, J., Yuan, S., Xu, Q., & Kod, M. (2016). A Novel Six-band Dual CP Rectenna Using Improved Impedance Matching Technique for Ambient RF Energy Harvesting. *IEEE Transactions on Antennas and Propagation*, *64*(7), 3160–3171. doi:10.1109/TAP.2016.2565697

Surface Mount Mixer and Detector Schottky Diodes, Data Sheet. (2013). Woburn, MA: Skyworks Solutions, Inc.

Wang, X., Zhang, L., Xu, Y., Bai, Y. F., Liu, C., & Shi, X.-W. (2013). A tri-band impedance transformer using stubbed coupling line. *Progress in Electromagnetics Research*, *141*, 33–45. doi:10.2528/PIER13042907

Wu, T., Arefin, M. S., Redouté, J., & Yuce, M. R. (2017). Flexible wearable sensor nodes with solar energy harvesting. *2017 39th Annual International Conference of the IEEE Engineering in Medicine and Biology Society (EMBC)*, 3273-3276. 10.1109/EMBC.2017.8037555

Chapter 4
Design of Slotted Hexagonal Wearable Textile Antenna Using Flexible Substrate

Lalita Kumari
CBS College of Engineering and Management, India

Lalit Kaushal
CBS College of Engineering and Management, India

Deepak Kumar
CBS College of Engineering and Management, India

ABSTRACT

In this chapter, a dual wideband textile antenna is proposed for WLAN and WiMax application. For antenna to be wearable, jeans material is used as a substrate to make ground plane, and copper tape is used to make patch of the anticipated antenna. The proposed antenna shows dual band performance with bandwidth of 82.48% covering 1.456 GHz to 3.5 GHz and 13.39% covering 4.32 GHz to 4.94 GHz. The simulated results like reflection coefficient, directivity, and radiation characteristics have been studied and analyzed.

INTRODUCTION

Textile Antennas are invaluable as a result of their compelling cost and straightforward acknowledgment process. There is a great deal of strategies to build the bandwidth, by expanding the thickness of substrate, utilization of

DOI: 10.4018/978-1-5225-9683-7.ch004

low dielectric substrate, utilizing various nourishing systems and by taking fractional ground (Klemm, Locher, and Troster, 2004). The proposed antenna configuration is utilized jeans which improves the bandwidth of antenna (Singh, Singh, and Singh, 2015; Bappadittya, Bhatterchya and Choudhury, 2013; Xu and Li, 2012).

Another period for the attire industry is the reconciliation of gadgets into textile messengers. The article of clothing of things to come won't just ensure the human body against the limits of nature yet additionally give data about the wearer's condition of wellbeing and condition (Bappadittya, Bhatterchya and Choudhury, 2013) appeared in Figure 1. This work not just goes for growing such wearable textile systems mostly for expert firemen and crisis calamities work force yet in addition for regular citizen casualties of common and different catastrophes. The fireman's inward and external article of clothing is being outfitted with an assortment of sensors (Bappadittya, Bhatterchya and Choudhury, 2013). This off-body correspondence requires the improvement of appropriate antenna that consolidates adaptability with heartiness and unwavering quality. An assortment of antennas for body-driven correspondence has been presented as of late (Grilo, and Correra, 2015; Rawat and Sharma, 2014; Chandran and Scanlon, 2010; Osman, Rahim, Samsuri, Zubir and Kamardin, 2011; Singh, Ali, Avub, and Singh, 2014). Because of the special shape and conservativeness, textile antennas have turned out to be most appropriate for coordination into articles of clothing (Srivastava, Singh, and Avub, 2015; Singh, Ali, Avub, and Singh, 2014; Singh, Singh, and Singh, 2015).

Radiating patch and ground of antenna is made of copper self sticky tape. Reproduction is finished by utilizing CST programming studio and gives the outcome, for example, reflection coefficient, addition and data transmission. The real advantages of the material radio wires are lightweight, low manufacture cost, low upkeep cost and hearty. Flexible antenna requires less space for establishment as these are basic and little in size. The main object is the hole for feed line which is put at the back of the ground plane (Srivastava, Singh, Ali, and Singh, 2013; Din, Chakrabarty, Ismail, Devi and Chen, 2012; Srivastava, Avub, and Singh, 2012).

ANTENNA DESIGN

The textile antenna substrate is made using a 1 mm thick jeans textile. Jeans is chosen in this work based on its cost effectiveness, ease of accessibility and ease of fabrication in comparison to other textile materials. The copper tape is used to form the ground plane and radiating elements of the anticipated antenna. The antenna results were obtained through CST software (Singh, Singh, Saxena [16-20]. Figure 1 illustrates the design of antenna consist of partial ground, made of copper adhesive tape of thickness 0.038mm and substrate is made of jeans with thickness 1mm.The proposed antenna provides triple band and hence suitable for wireless communication system.

The length and width of ground plane are 17 mm and 60 mm. The general permittivity of jeans substrate is 1.7 with tangent digression of 0.025.The radius of hexagonal patch is 25 mm with two square slots with measurement 10x10 mm and the dimension of microstrip feed line is 30.9 x2 mm. The radius of circle slot is 14 mm.

Figure 1. Configuration of Antenna

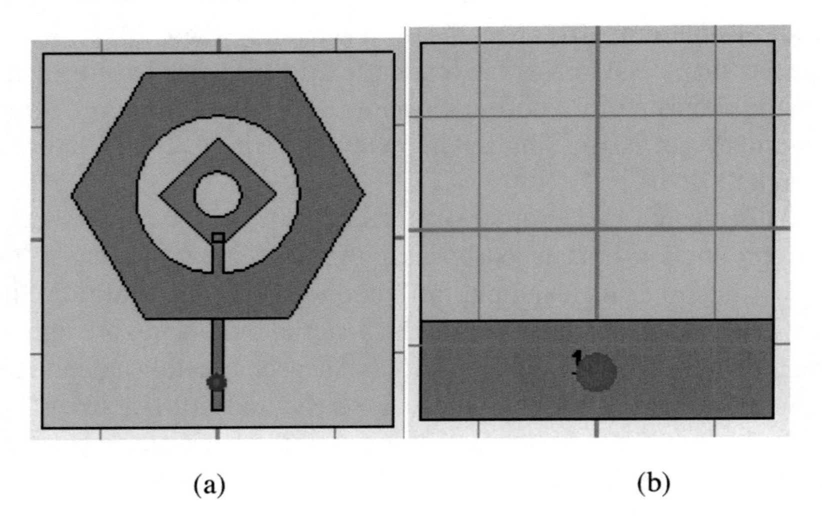

(a) (b)

RESULTS AND DISCUSSIONS

This work not just goes for growing such wearable textile systems mostly for expert firemen and crisis calamities work force yet in addition for regular citizen casualties of common and different catastrophes. Figure 2 shows return loss Vs frequency graph at resonant frequency 1.708 GHz. The 3D characteristics of proposed antenna are shown in Figure 3 which shows directivity 2.682 dBi at 1.708 GHz.

Figure 2. Reflection coefficient Vs frequency

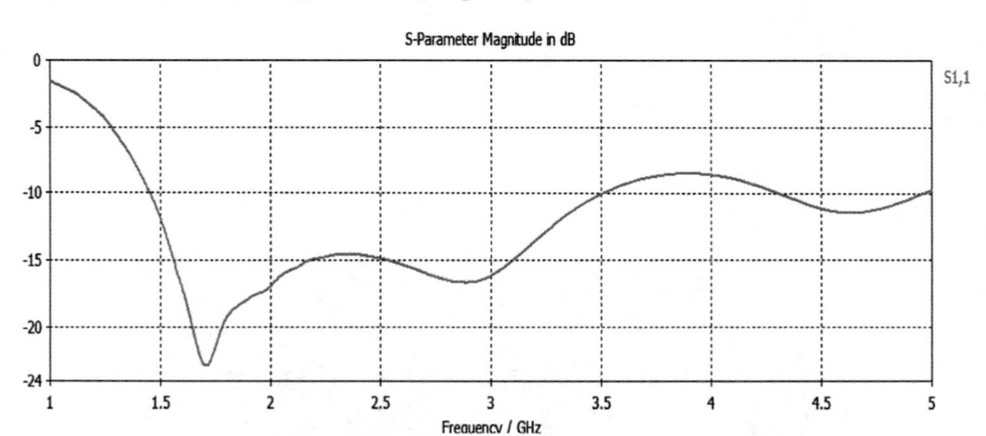

Figure 3a. 3-D Radiation pattern: Directivity at 1.708 GHz

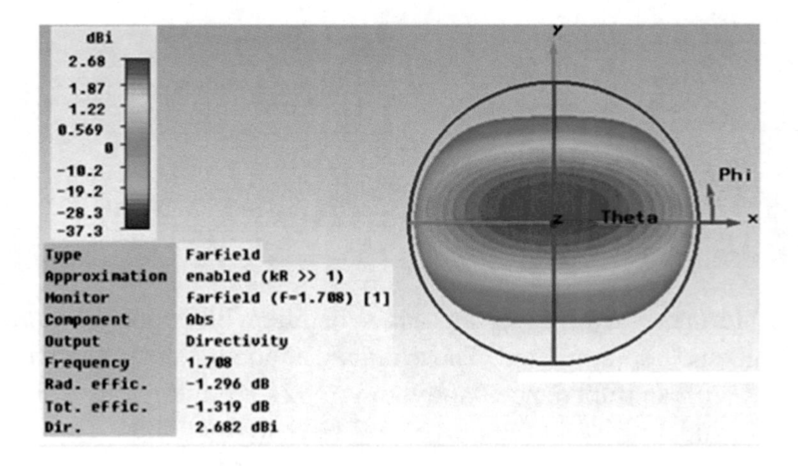

Figure 3b. 3-D Radiation pattern: Left polarization at 1.708 GHz

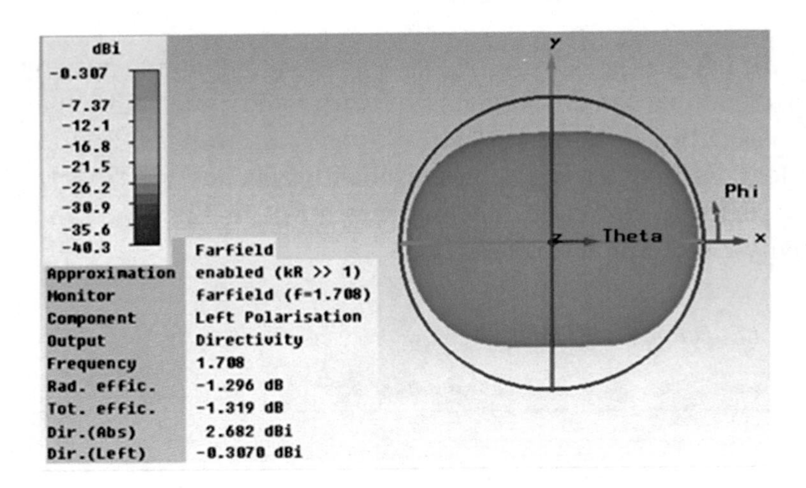

Figure 3c. 3-D Radiation pattern: Right polarization at 1.708 GHz

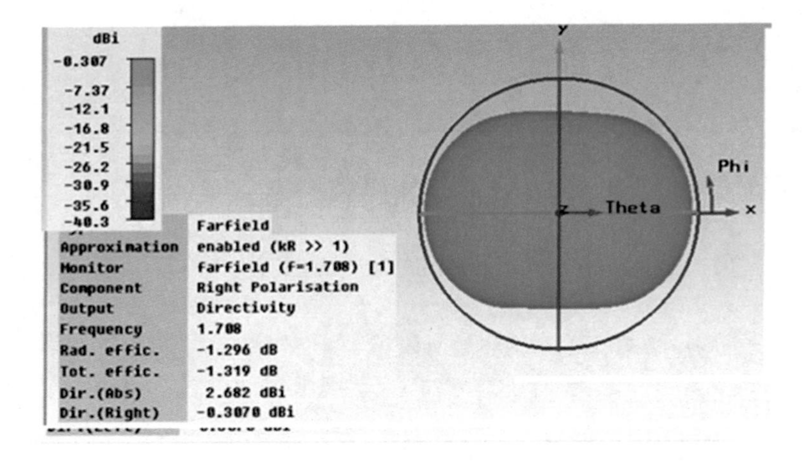

CONCLUSION

The flexible antenna utilizing jeans as a substrate with double band attributes is exhibited and contemplated. The arranged adaptable antenna gives double wide band with an impedance bandwidth of 82.48% in the range 1.456 GHz to 3.5 GHz and 13.39% in the range 4.32 GHz to 4.94 GHz. The antenna is best appropriate for various remote correspondence framework, for example,

WLAN (2.40GHz – 2.48GHz), Bluetooth (2.45 GHz) and WiMAX (2.495GHz-2.695GHz) applications.

REFERENCES

Chandran, A. R., & Scanlon, W. G. (2010). Dual-band low probe antennas for body-centric communications. *2010 International Workshop on Antenna Technology (IWAT)*, 1-3.

Din, N. M., Chakrabarty, C. K., Bin Ismail, A., Devi, K. K. A., & Chen, W. Y. (2012). Design of RF Energy Harvesting System For Energizing Low Power Devices. *Progress in Electromagnetics Research*, *132*, 49–69. doi:10.2528/PIER12072002

Grilo, M., & Correra, F. S. (2015). Rectangular patch antenna on textile substrate fed by proximity coupling. *Journal of Microwaves, Optoelectronics and Electromagnetic Applications*, *14*, 103–112.

Klemm, M., Locher, I., & Tröster, G. (2004). A Novel Circularly Polarized Textile Antenna for Wearable Applications. Proc. 7th European Microwave Week, 137-140.

Mai, A. R. (2011). Design, implementation and performance of ultra-wideband textile antenna. *Progress In Electromagnetics Research B*, *27*, 307–325. doi:10.2528/PIERB10102005

Nikhil Singh, A. K. (2015). Design and performance of wearable ultra wide band textile antenna for medical applications. *Microwave and Optical Technology Letters*, *57*(7), 1553–1557. doi:10.1002/mop.29131

Rawat, S., & Sharma, K. K. (2014). A compact broadband microstrip patch antenna with defected ground structure for C-band applications. *Central European Journal of Engineering*, *4*(3), 287–292.

Roy, Bhatterchya, & Choudhury. (2013). Characterization of textile substrate to design a textile antenna. *2013 International Conference on Microwave and Photonics (ICMAP)*.

Sharma, P., Yadav, A., & Singh, V. K. (2018). Design of Circularly Polarized Antenna with Different Iterations for UWB Applications. In A. Kalam, S. Das, & K. Sharma (Eds.), *Advances in Electronics, Communication and Computing. Lecture Notes in Electrical Engineering* (Vol. 443, pp. 441–447). Springer. doi:10.1007/978-981-10-4765-7_47

Sharma, P., Yadav, A., & Singh, V. K. (2018). Design of Circularly Polarized Antenna with Different Iterations for UWB Applications. In A. Kalam, S. Das, & K. Sharma (Eds.), *Advances in Electronics, Communication and Computing. Lecture Notes in Electrical Engineering* (Vol. 443, pp. 441–447). Springer. doi:10.1007/978-981-10-4765-7_47

Singh. (2014). *A wide band Compact Microstrip Antenna for GPS/DCS/PCS/ WLAN Applications. In Intelligent Computing, Networking, and Informatics* (Vol. 243, pp. 1107–1113). Springer.

Singh, N., Singh, A. K., & Singh, V. K. (2015). Design and Performance of Wearable Ultra Wide Band Textile Antenna for Medical Applications. Microwave and Optical Technology Letters, 57(7), 1553-1557.

Singh, N. K., Sharma, N., Ali, Z., Singh, V. K., & Bhoi, A. K. (2018). Inset Fed Circular Microstrip Antenna with Defected Ground. In A. Kalam, S. Das, & K. Sharma (Eds.), *Advances in Electronics, Communication and Computing. Lecture Notes in Electrical Engineering* (Vol. 443, pp. 605–611). Singapore: Springer. doi:10.1007/978-981-10-4765-7_63

Singh, N. K., Singh, V. K., Saxena, A., Bhoi, A. K., Garg, A., & Sherpa, K. S. (2018). A Compact Slotted Textile Patch Antenna for Ultra-wide Band Application. In A. Kalam, S. Das, & K. Sharma (Eds.), *Advances in Electronics, Communication and Computing. Lecture Notes in Electrical Engineering* (Vol. 443, pp. 53–59). Springer. doi:10.1007/978-981-10-4765-7_6

Singh, R., Singh, V. K., & Khanna, P. (2018). A Compact CPW-Fed Defected Ground Microstrip Antenna for Ku Band Application. In A. Kalam, S. Das, & K. Sharma (Eds.), *Advances in Electronics, Communication and Computing. Lecture Notes in Electrical Engineering* (Vol. 443, pp. 231–237). Springer. doi:10.1007/978-981-10-4765-7_24

Singh, V. K., Ali, Z., Ayub, S., & Singh, A. K. (2014). Bandwidth Optimization of compact Microstrip Antenna for PCS/DCS/Bluetooth Application. *Central European Journal of Engineering, Springer*, *4*(3), 281–286.

Srivastava, R. (2015). *Comparative Analysis and Bandwidth Enhancement with Direct Coupled C Slotted Microstrip Antenna for Dual Wide Band Applications. In Frontiers of Intelligent Computing: Theory and Applications* (Vol. 328, pp. 449–455). Springer.

Srivastava, R., Ayub, S., & Singh, V. K. (2014). Dual Band Rectangular and Circular Slot Loaded Microstrip Antenna for WLAN/GPS/WiMax Applications. *IEEE Conference on Communication Systems and Network Technologies (CSNT-2014)*, 45 - 48.

Srivastava, S., Singh, V. K., Ali, Z., & Singh, A. K. (2013). Duo Triangle Shaped Microstrip Patch Antenna Analysis for WiMAX lower band Application. *Procedia Technology, 10*, 554-563.

Xu, Joshua, & Li. (2012). A Dual band Microstrip Antenna for Wearable Application. *IEEE Conference Proceeding*, 109-112.

Chapter 5
Development of Novel Design to Enhance the Characteristics of Flexible Antenna

Neha Nigam
S. R. Group of Institution Jhansi, India

Vinod Kumar Singh
https://orcid.org/0000-0002-0671-0631
S. R. Group of Institution Jhansi, India

ABSTRACT

This chapter proposed triple band novel geometry and enhanced characteristics of flexible textile antenna. The proposed radio wire indicates wideband execution with wide data transfer capacity of 20.50% covering the recurrence scope of 6.3039 GHz to 7.7445 GHz, 11.57% covering the recurrence scope of 9.0694 GHz to 10.184 GHz, and 8.23% in the recurrence scope of 12.497 GHz to 13.57 GHz. In this chapter, reenacted outcomes like return loss, directivity, and radiation characteristics have been contemplated.

1. INTRODUCTION

The fast advancement of remote power transfer has expanded the interest for textile antenna with high gain and improved band attributes. Textile Antenna is exceptionally invaluable due to their successful cost, small profile, low mass and basic acknowledgment process. There are a great deal of strategies to expand the bandwidth of Antenna, by expanding the tallness of substrate,

DOI: 10.4018/978-1-5225-9683-7.ch005

utilization of low dielectric substrate, utilizing various sustaining methods and by taking defected ground [1-5]. The proposed antenna configuration is utilized jeans as a substrate whose relative permittivity is around 1.7. Because of low dielectric constant the attenuation are exceptionally less and it likewise upgrades the data bandwidth of Antenna [6-9].The initial phase in the plan of an antenna comprises of picking suitable material for the substrate and the directing part. Decision of choice of substrate the most appropriate substrate for an antenna involves prime significance. This is on the grounds that numerous restrictions of the microstrip wearable antenna, for example, exceptional reflection coefficient, and low gain of antenna [10-11]. The substrate properties, for example, its dielectric constant and loss tangent pronouncedly affect the antenna attributes.

A portion of the basic characteristics that are to be dealt with while choosing a dielectric are homogeneity and dampness retention. Consequently, we pick an uncommon sort of fabric (Jeans) with a thickness of 1 mm substrate. Low ε_r of the reception apparatus substrate permits the structure of textile antenna with a vast gain and a high effectiveness [1]. After simulation it has been received the outcome, for example, reflection coefficient, gain and data transfer capacity. The real advantages of the antenna are lightweight, low creation cost, low upkeep cost and vigorous. Flexible antenna requires not as much of space for establishment as these are basic and little in size. The main space these need is the hole for feed line which is set at the back of the ground plane [12-13].

2. ANTENNA DESIGN PROCEDURE

This paper presents the design of proposed antenna geometry and analyzed on simulation software. The antenna geometry consists of conducting patch, dielectric substrate material and a partial conducting patch used as ground plane. The copper is used to form the ground plane and radiating elements of the anticipated antenna. CST is used to simulate the anticipated design. Fig.1 illustrates the design of flexible antenna consist of partial ground, made of copper adhesive tape of thickness 0.038mm and substrate is made of jeans with thickness 1mm.The proposed antenna provides triple band and hence suitable for wireless communication system. The dimensions of ground are 30 mm and 86 mm. The general permittivity of jeans substrate is 1.7 with tangent digression of 0.025.The radius of patch is 14 mm with two square slots with measurement 5x5 mm and the width of feed line is 2 mm.

Figure 1. Novel geometry of flexible antenna

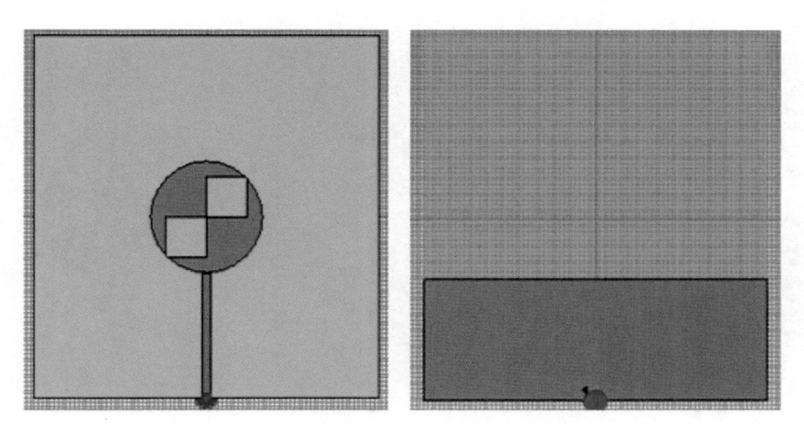

3. RESULTS AND DISCUSSIONS

The anticipated antenna features has been analyzed and studied. Fig.2 shows return loss Vs frequency plot with triple band at 6.8086GHz, 9.5952 GHz and 13.044GHz respectively. The 3D characteristics of proposed antenna are shown in Fig.3 which shows directivity 3.953dBi, 5.017dBi and 6.437dBi. Fig.4 explains the polar characteristics of the anticipated antenna at resonant frequencies 6.8086GHz, 9.5952GHz and 13.044 GHz respectively.

Figure 2. Return loss Vs frequency simulated graph

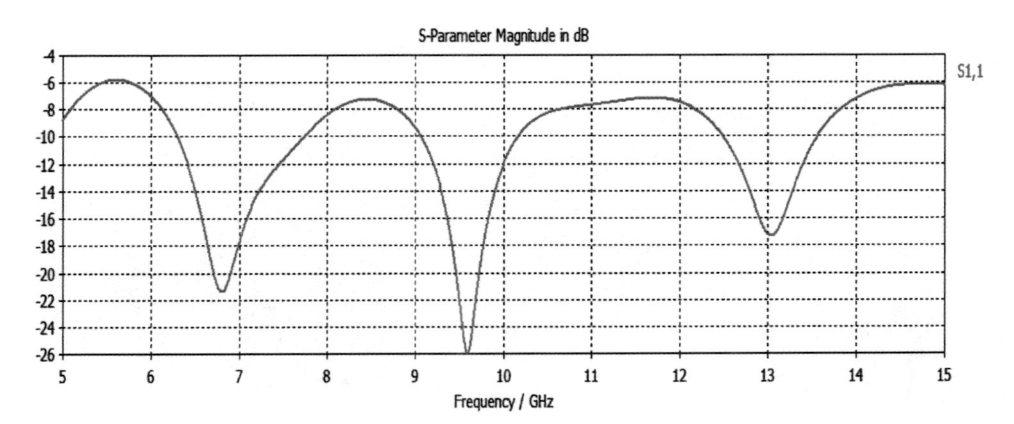

Figure 3. Fairfield at (i) 6.8086GHz (ii) 9.5952GHz (iii) 13.044 GHz

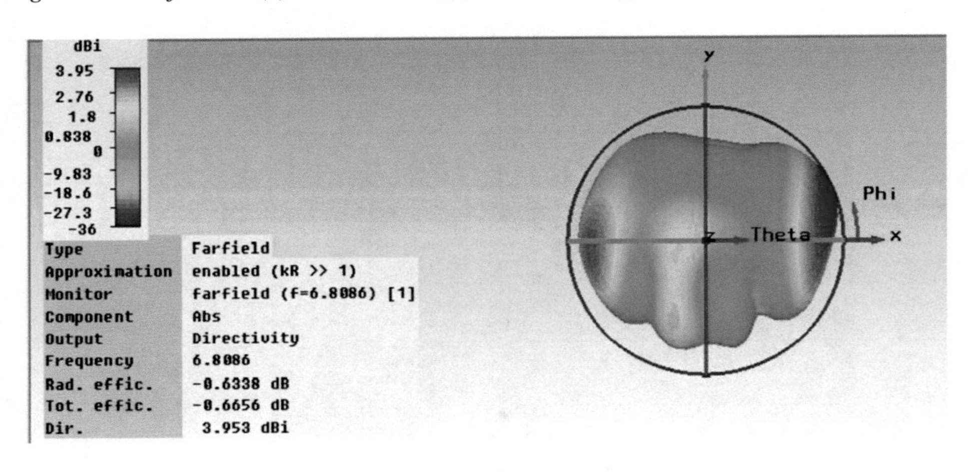

(i)

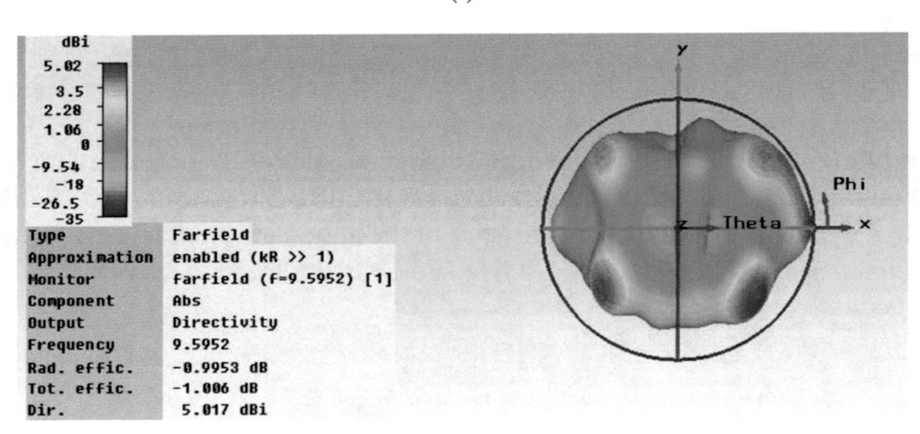

(ii)

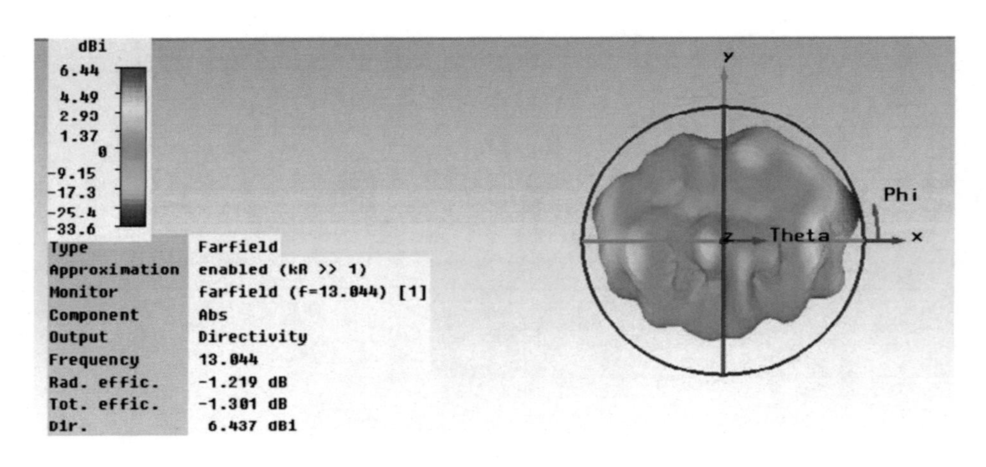

(iii)

Figure 4. Polar plot of at (i) 6.8086GHz (ii) 9.5952GHz (iii) 13.044GHz

Directivity Abs (Phi=90)

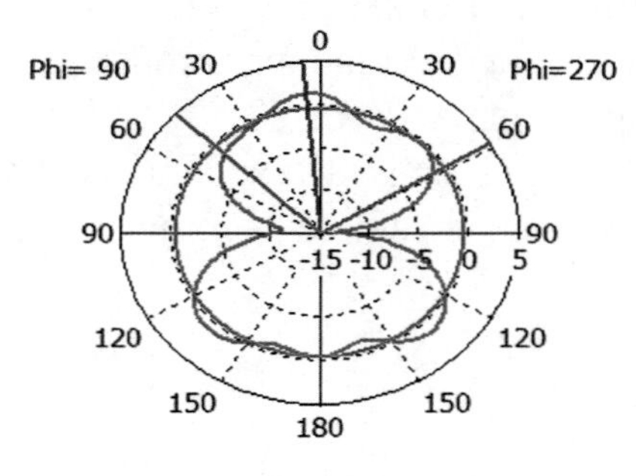

Theta / Degree vs. dBi

farfield (f=6.8086) [1]

Frequency = 6.8086
Main lobe magnitude = 1.3 dBi
Main lobe direction = 5.0 deg.
Angular width (3 dB) = 104.6 deg.
Side lobe level = -1.7 dB

(i)

Directivity Abs (Phi=90)

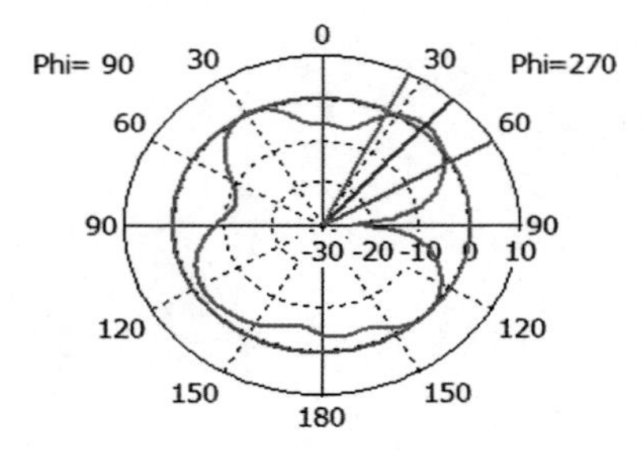

Theta / Degree vs. dBi

farfield (f=9.5952) [1]

Frequency = 9.5952
Main lobe magnitude = 1.7 dBi
Main lobe direction = 41.0 deg.
Angular width (3 dB) = 34.3 deg.
Side lobe level = -1.1 dB

(ii)

Directivity Abs (Phi=90)

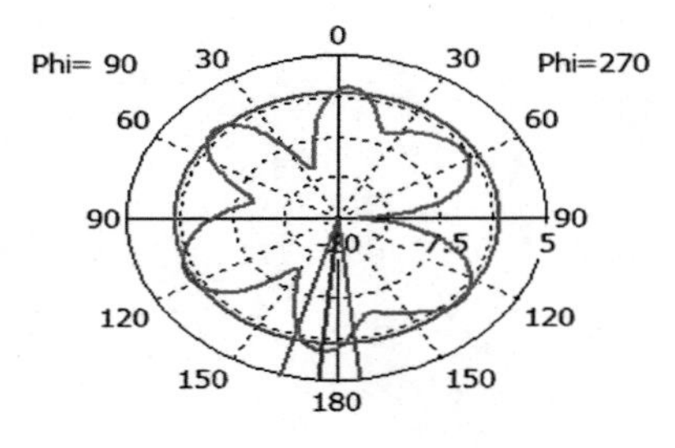

Theta / Degree vs. dBi

farfield (f=13.044) [1]

Frequency = 13.044
Main lobe magnitude = 0.5 dBi
Main lobe direction = 175.0 deg.
Angular width (3 dB) = 22.2 deg.
Side lobe level = -1.0 dB

(iii)

4. CONCLUSION

The jeans are adaptable material that is utilized as a substrate to plan antenna with triple band attributes is presented and examined. A planned antenna gives three distinctive wide bands with data transfer capacity of 20.50%, 11.57% and 8.23% with the gain of 3.93dBi, 5.017dBi and 6.436dBi separately. The anticipated antenna is best appropriate for various remote communication systems.

REFERENCES

Chandran, A. R., & Scanlon, W. G. (2010). Dual-band low probe antennas for body-centric communications. *2010 International Workshop on Antenna Technology (IWAT)*, 1-3.

Din, N. M., Chakrabarty, C. K., Bin Ismail, A., Devi, K. K. A., & Chen, W. Y. (2012). Design of RF Energy Harvesting System For Energizing Low Power Devices. *Progress in Electromagnetics Research*, *132*, 49–69. doi:10.2528/PIER12072002

Grilo, M., & Correra, F. S. (2015). Rectangular patch antenna on textile substrate fed by proximity coupling. *Journal of Microwaves, Optoelectronics and Electromagnetic Applications*, *14*, 103–112.

Klemm, M., Locher, I., & Tröster, G. (2004). A Novel Circularly Polarized Textile Antenna for Wearable Applications. Proc. of 7th European Microwave Week, 137-140.

Mai, A. R. (2011). Design, implementation and performance of ultra-wideband textile antenna. *Progress In Electromagnetics Research B*, *27*, 307–325. doi:10.2528/PIERB10102005

Nikhil Singh, A. K. (2015). Design and performance of wearable ultra wide band textile antenna for medical applications. *Microwave and Optical Technology Letters*, *57*(7), 1553–1557. doi:10.1002/mop.29131

Rawat, S., & Sharma, K. K. (2014). A compact broadband microstrip patch antenna with defected ground structure for C-band applications. *Central European Journal of Engineering*, *4*(3), 287–292.

Roy, Bhatterchya, & Choudhury. (2013). Characterization of textile substrate to design a textile antenna. *2013 International Conference on Microwave and Photonics (ICMAP)*.

Singh. (2014). *A wide band Compact Microstrip Antenna for GPS/DCS/PCS/WLAN Applications. In Intelligent Computing, Networking, and Informatics* (Vol. 243, pp. 1107–1113). Springer.

Singh, N., Singh, A. K., & Singh, V. K. (2015). Design and Performance of Wearable Ultra Wide Band Textile Antenna for Medical Applications. Microwave and Optical Technology Letters, 57(7), 1553-1557.

Singh, V. K., Ali, Z., Ayub, S., & Singh, A. K. (2014). Bandwidth Optimization of compact Microstrip Antenna for PCS/DCS/Bluetooth Application. *Central European Journal of Engineering, Springer*, *4*(3), 281–286.

Srivastava, R. (2015). *Comparative Analysis and Bandwidth Enhancement with Direct Coupled C Slotted Microstrip Antenna for Dual Wide Band Applications. In Frontiers of Intelligent Computing: Theory and Applications* (Vol. 328, pp. 449–455). Springer.

Srivastava, R., Singh, V. K., & Ayub, S. (2015). Comparative Analysis and Bandwidth Enhancement with Direct Coupled C Slotted Microstrip Antenna for Dual Wide Band Applications. Advances in Intelligent Systems and Computing, 328, 449-455.

Srivastava, S., Singh, V. K., Ali, Z., & Singh, A. K. (2013). Duo Triangle Shaped Microstrip Patch Antenna Analysis for WiMAX lower band Application. *Procedia Technology, 10*, 554-563.

Xu, Joshua, & Li. (2012). A Dual band Microstrip Antenna for Wearable Application. *IEEE (ISAPE) Conference Proceeding*, 109-112.

Chapter 6
Efficient Rectenna Circuit for Wireless Power Transmission

Anurag Saxena
S. R. Group of Institutions Jhansi, India

Paras Raizada
S. R. Group of Institutions Jhansi, India

Lok Prakash Gautam
S. R. Group of Institutions Jhansi, India

Bharat Bhushan Khare
https://orcid.org/0000-0001-8755-9808
UIT RGPV Bhopal, India

ABSTRACT

Wireless power transmission is the transmission of electrical energy without using any conductor or wire. It is useful to transfer electrical energy to those places where it is hard to transmit energy using conventional wires. In this chapter, the authors designed and implemented a wireless power transfer system using the basics of radio frequency energy harvesting. Numerical data are presented for power transfer efficiency of rectenna. From the simulated results, it is clear that the anticipated antenna has single band having resonant frequency 2.1 GHz. The anticipated antenna has impedance bandwidth of 62.29% for single band. The rectenna has maximum efficiency of 60% at 2.1 GHz. The maximum voltage obtained by DC-DC converter is 4V at resonant frequency.

DOI: 10.4018/978-1-5225-9683-7.ch006

INTRODUCTION

For converting electromagnetic energy (AC) into direct current (DC) power or electricity a special type of antenna i.e. rectifying antenna is used for the task. Rectenna transmit the energy or power by radio waves in wireless power transmissions. A Rectenna element consists of an antenna with an RF diode connected across the elements (Tesla, 1905, pp. 21-24; Crawford, 2005; Naresh and Singh, 2017a). The diode works on forward bias to rectify the AC energy induced in the antenna by the microwaves, to generate DC power, in which power was delivered at the load. Schottky diodes are frequently used because they have the lowest voltage fall and maximum speed and therefore have the lowest energy losses due to transfer and switching (Naresh and Singh, 2017b; Dickson, 2013, pp. 36-47; Naresh, Singh, Bhargavi, n.d., --. 131-138; Parviz, 2009, pp. 36-41).

Recently energy harvesting was the main focus of the research community. There are various sources of power that energy harvesting can gain from. Micro strip patch antennas have been an vast topic for study due to its better configuration which comprises of fine structure, smaller size and lesser weight and more economical. Wireless power transmission holds reliable criteria for future work in generating the electricity for charging mobile wirelessly, as there is no such requirement in placing the cell phones at a very shorter distance to the sockets due to the shorter length of the cable as compared to those of area covered by the wireless field (Warneke, 2001, pp. 44-51; Glaser, 1968, pp. 857-861; Shinohara, 200; Dobkin and Weigand, 2007, pp. 170-190; Brown, 1964, pp. 8-17).

RECTENNA CIRCUIT

High density integration technologies have provided immigration from mobile to wearable in information communication system in recent times. In this paper, an anticipated antenna is simulated using slotting techniques on a PCB. They are mostly used at microwave frequencies. A microstrip patch antenna can be a type of metal foil of different design on the face of a PCB, with a ground plane on the other side of the PCB. Microstrip patch antennas have turn into well-liked in modern times by reason of their low weight low shape conformability, easy and cheap realization. The simulations were carried out using IE3D software and design parameters are shown in Table 1. Figure 1 shows the top view of proposed antenna Geometry.

$$\lambda = \frac{c}{f} \tag{1}$$

The efficiency wireless power system is measure as the power (in %) i.e. transferred from the power source to the received power. The efficiency of harvesting circuit can be measure by eq.(2).

$$\eta = \frac{P_{out}}{P_{in}} \times 100 \tag{2}$$

Pout = DC Power output
Pin = DC Power input

Table 1. Dimension of microstrip antenna

S.No	Parameter	Value
1	Substrate thickness (mm)	1.6
2	Relative permittivity	4.4
3	Length of the patch (mm)	33.72
4	Width of the patch (mm)	43.47

Figure 1. Top view of proposed antenna geometry

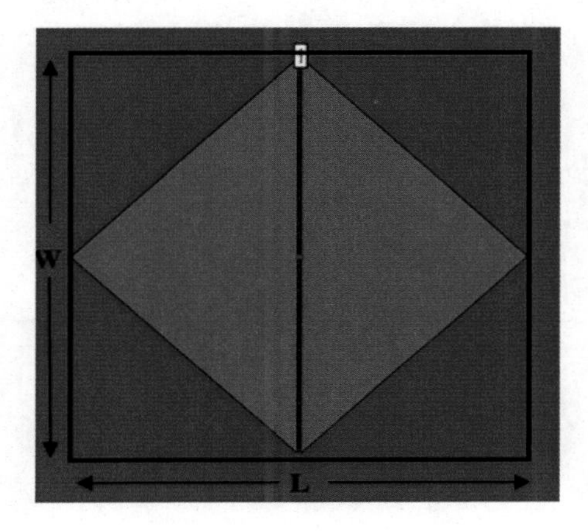

RECTIFICATION CIRCUIT

With the second order low pass filter, rectifying circuit is designed which results to the impedance matching of antenna and diode. Matching is essential so that harvesting efficiency will reduce by power. For Radio Frequency to DC conversion Schottky diode is used with threshold voltage of 150 mV. The input impedance of the diode is a dynamic variable dependent on input RF power. LC filter elements are calculated separately for dual band frequencies. If we use rectenna circuit for sensor application, the main failure is the ripples which are generated at the output. Since these are very much sensitive devices, the effect is more. Therefore filters are necessary to remove the ripples so that flat DC current can be passed from the load. This will remove the AC components at the output of proposed rectenna. The proposed low power rectenna electrical circuit is shown in Figure 2. The circuit elements for 2.1 GHz are L1= 3.79nH, C_1=0.76pF, L_2= 16.52nH and C2= 0.33pF.

The main function of half wave diode rectifier is to convert RF signal to DC signal and also to rectify the received microwave power into DC power at very low voltage. To boost up the output voltage of DC voltage such that it can drive the microelectronic devices. Therefore to step-up the rectified DC voltage, DC to DC converter is used.

RESULTS AND DISCUSSION

Rectenna has been used to convert wireless input energy into output DC energy. The conjugate tuning of a rectifying circuit directly to a microstrip antenna, a matching and filtering system among the antenna and rectifying circuit can be avoided. Since it is not easily accessible for the input power coupled into the rectifying circuit therefore a relative measurement was performed.

Figure 2. Diode rectifier circuit

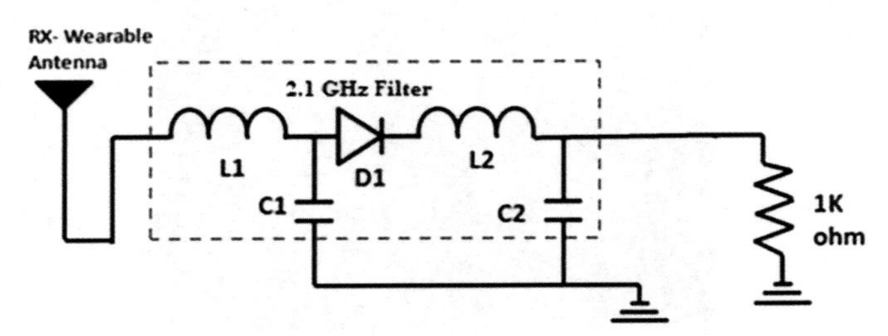

Figure 3. Frequency response Vs reflection coefficient plot of anticipated microstrip antenna

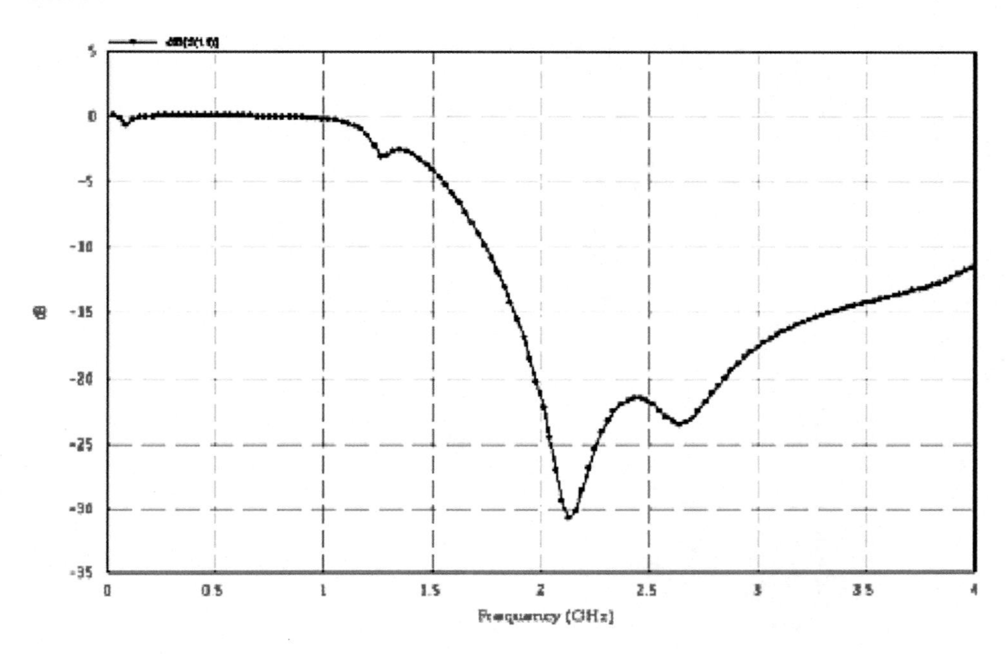

This paper describes the analysis of the microstrip patch antenna design to obtain better results. The substrate dimension is 33.72 x 43.47 mm. Figure 3 represents the simulated return loss vs. frequency response plot of presented antenna. The simulated graph of reflection coefficient at 2.1 GHz is shown in Figure 3. Figure 4 shows the output voltage vs. power of Schottky diode. Figure 5 represents the Schottky diode efficiency vs. power input.

The distance (Dr) between the transmitting horn antenna with the gain of G_t = 11dBi and the rectenna is one meter. For finding out the micro power Friis transmission Equation is used which is available at rectenna terminals by using eq. (3).

$$P_r = P_t G_t G_r \left(\frac{C}{4\pi D_r f_o} \right)^2 \quad (3)$$

Where P_t is the transmitting power at a given field strength E (mV/m); G_r is the receiving antenna gain (3.0 dBi).The term C and f_0 are the velocity of light and frequency of the microwave. The output DC voltage (VoutDC) and overall efficiency η of the rectenna against power density are calculated by eq. (4).

$$\eta = \frac{P_{outDC}}{P_{rX}} = \frac{{V^2}_{outDC} / R_L}{P_r} \tag{4}$$

Figure 4. Output voltage vs power

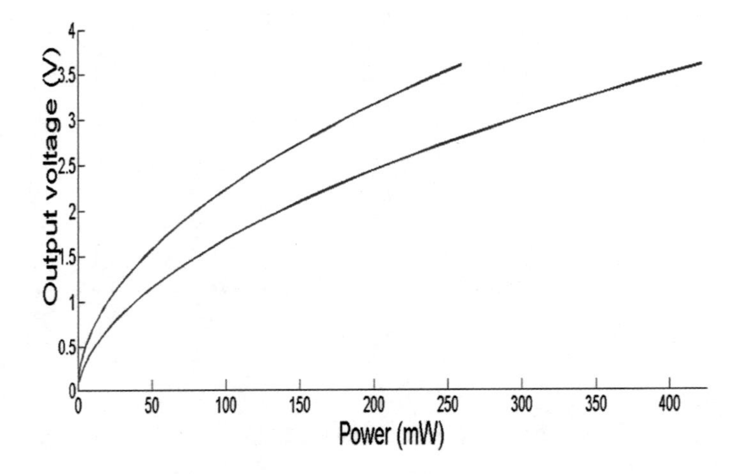

Figure 5. Schottky diode efficiency vs power input

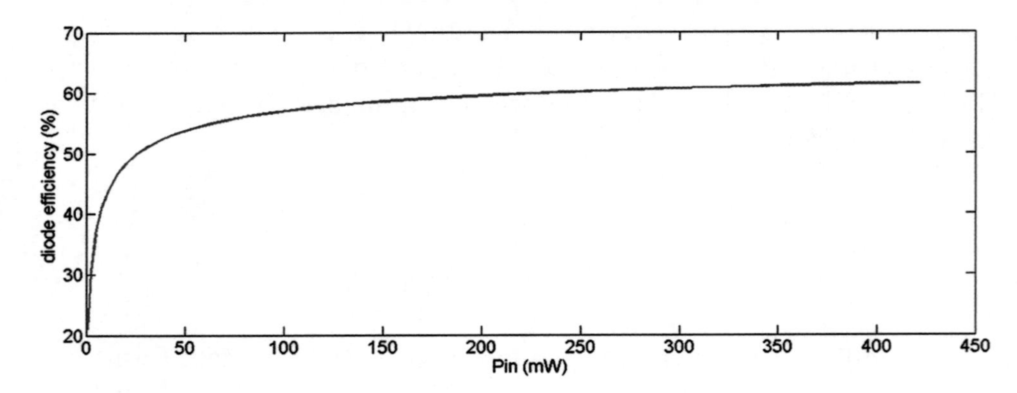

CONCLUSION

This paper explains the concept of energy harvesting and also its potential function in wireless system. From the simulated results the anticipated antenna has single band having resonant frequency 2.1 GHz. Proposed rectenna is tested for different micro power levels as -20 dBm to 15 dBm. The anticipated antenna has impedance bandwidth of 62.29% for single band. The rectenna has maximum efficiency of 60% at 2.1 GHz. The maximum voltage obtained by DC-DC converter is 4V at resonant frequency.

REFERENCES

Brown, W. (1964). Experiments in the transportation of energy by microwave beam. *IRE Int. Convention Record, 12*(2), 8–17. 10.1109/IRECON.1964.1147324

Crawford, G. (2005). *Flexible Flat Panel Displays*. New York: Wiley. doi:10.1002/0470870508

Dickson, R. M. (2013). Power in the sky: Requirements for microwave wireless power beamers for powering high-altitude platforms. *IEEE Microwave Magazine, 14*(2), 36–47. doi:10.1109/MMM.2012.2234632

Dobkin, D. M., & Weigand, S. M. (2007). UHF RFID and tag antenna scattering, Part I: Experimental results. *Microwave J. Euro-Global Ed., 49*(5), 170–190.

Glaser, P. E. (1968). Power from the sun: Its future. *Science, 162*(3856), 857–861. doi:10.1126cience.162.3856.857 PMID:17769070

Naresh & Singh. (2017). Dual band RF Energy Harvester for Wearable Electronic Technology. *3rd International Conference on Advances in Electrical, Electronics, Information, Communication and Bio-Informatics (AEEICB-17)*. 10.1109/AEEICB.2017.7972428

Naresh & Singh. (2017). 4.65 GHz Wearable Rectenna for low power Wireless applications. *International Conference on Electrical, Computer and Communication Technologies (CECCT-2017).*

Naresh, B., Singh, V. K., & Bhargavi, V. Low Power Circularly Polarized Wearable Rectenna for RF Energy Harvesting. In A. Garg, A. Bhoi, P. Sanjeevikumar, & K. Kamani (Eds.), *Advances in Power Systems and Energy Management. Lecture Notes in Electrical Engineering* (Vol. 436, pp. 131–138). Singapore: Springer. doi:10.1007/978-981-10-4394-9_13

Naresh, B., Singh, V. K., Bhargavi, V., Garg, A., & Bhoi, A. K. Dual-Band Wearable Rectenna for Low-Power RF Energy Harvesting. In A. Garg, A. Bhoi, P. Sanjeevikumar, & K. Kamani (Eds.), *Advances in Power Systems and Energy Management. Lecture Notes in Electrical Engineering* (Vol. 436, pp. 13–21). Singapore: Springer. doi:10.1007/978-981-10-4394-9_2

Parviz, B. (2009, September). Augmented reality in a contact lens. *IEEE Spectrum, 46*(9), 36–41. doi:10.1109/MSPEC.2009.5210042

Shinohara, N. (2000). *Wireless power transmission for solar power satellite (SPS).* Space Solar Power Institute, Tech. Rep.

Tesla, N. (1905, January). The transmission of electrical energy without wires as a means for furthering peace. *Elect. World Eng.*, 21–24.

Warneke, B., Last, M., Liebowitz, B., & Pister, K. (2001). Smart dust: Communicating with a cubic-millimeter computer. *Computer, 34*(1), 44–51. doi:10.1109/2.895117

Chapter 7
Multi-Band Rectangular Zig-Zag-Shaped Microstrip Patch Antenna for Wireless Applications

Pranay Yadav
https://orcid.org/0000-0002-9368-8398
Ultra-Light Technology Bhopal, India

Bharat Bhushan Khare
https://orcid.org/0000-0001-8755-9808
UIT RGPV Bhopal, India

Sudesh Gupta
Technocrats Institute of Technology Bhopal, India

Yash Kumar Kshirsagar
Bhopal School of Social Sciences, India

Swati Jain
UIT RGPV Bhopal, India

ABSTRACT

In the era of fifth generation communication system, multi-band patch antenna is the key element of this system. Most of the wireless communication devices work in the range of 1 to 6 GHz such as wireless fidelity (Wi-Fi), Wi-Max, and wireless local area network. In this research work, a modified zig-zag-shaped multiband patch antenna with cross-cut set that covers 1 to 6 GHz range is designed. Proposed patch antenna shows five different bands at

DOI: 10.4018/978-1-5225-9683-7.ch007

different resonant frequencies 1.4, 2.3, 2.5, 3.42, and 4.16 all in GHz. The proposed zig-zag-shaped patch antenna also calculates the radiation pattern and specific absorption rate (SAR).

INTRODUCTION

In this paper the knowledge of the wireless communication, the conventional Microstrip patch antenna, dielectric resonator antenna and the introduction of new advanced technology that is electromagnetic band gap structures are defined. The crucial component of a wireless network is the 'Antenna' so without appropriate design of the antenna, the signal produced by the RF system will not be transmitted properly and not any proper interfered signal can be detected at the receiver end (Luk, 2003 & Yang and Rahmat-Samji, 2003 & Khalily et. al, 2014 & Mu'ath, Denidni, and Sebak, 2012). Various types of antenna have been made to provide various applications and suitable for particular requirements.

The microstrip antenna has been considered to be the most ingenious field in the engineering of antenna having the properties like low material cost and the easy simulation. So the frequency range is also increased slightly in the some millimeter region (approx 100-300 GHz) that can be utilized (Al-Hasan, Denidni, and Sebak 2012 & Yadav and Singh, 2019 & Bhardwaj and Kumar, 2012 & Ryu and Kishk, 2010 & Coulibaly and Denidni, 2009). The conduction losses in metallic antennas have expanded to a level that affects the relevant operation of the systems. So an idea to use dielectric material as a radiator was perceived. These antennas are used in space applications, government and commercial applications. They include radiating patch of metallic material on substrate with ground structure on its back (Ge and Esselle, 2009 & Aras et al, 2008 & El-Deen et al, 2006).

This paper covered the basic principle of the design of dielectric resonator antenna with the help of mushroom-like electromagnetic band gap structure. The design and simulation of triple-band rectangular zigzag micro-strip patch antenna having Defected Ground Structure (DGS) for improved bandwidth for WLAN AND DBS applications. The fabrication and testing of Triple-Band Rectangular Zigzag microstrip patch antenna has been done (Li and Sun, 2005 & Mittal, Khanna, and Kaur, 2015 & Singh, Dhupkariva, and Bangari, 2017 & Kaur, Khanna, and Kartikeyan, 2014). In this micro-strip patch antenna, slotting and different feeding technique is also used to enhance the gain.

Micro-strip Antenna has become popular day by day because the fabrication of such kinds of antenna is very easy. Microstrip Patch Antenna is designed for Giga hertz frequency range where this frequency range accommodate at the various band in the 1.4, 2.3, 2.5, 3.42,4.16 all are in GHz range frequency spectrum and Radiate Wave (Liu et. al, 2011 & Pei et. al, 2011 & Dong, Toyao, and Itoh, 2011 & Reddy and Vakula, 2014). Microstrip Patch Antennas have Gaining importance in the applications of Wireless Local Area networks (WLAN). The simulated results such as Return Loss S_{11}, VSWR, Gain, and Radiation Pattern, Vector diagram of electric field and Mesh field and one of the important parameter that is Specific absorption rate (SAR) are presented in this proposed design for Zigzag Shaped slotted patch with DGS. New proposed micro-strip antenna has enhanced gain and radiation pattern. This explains the designing process of Multi-Band Rectangular Zigzag Microstrip Patch Antenna with Defected Ground Structure (DGS) (Chakraborty et. al, 2017 and Singh et. al, 2013) with some improvement in bandwidth for WLAN Applications. The bandwidth can be improved using slots on the radiating patch, introducing defects in the grounded structure, impedance matching.

DESIGN OF PROPOSED ZIG ZAG ANTENNA

The optimum parameters are determined by many simulated results to achieve the design of proposed antenna. The design is simulated on high frequency

Figure 1a. Design of rectangular zigzag shaped microstrip patch

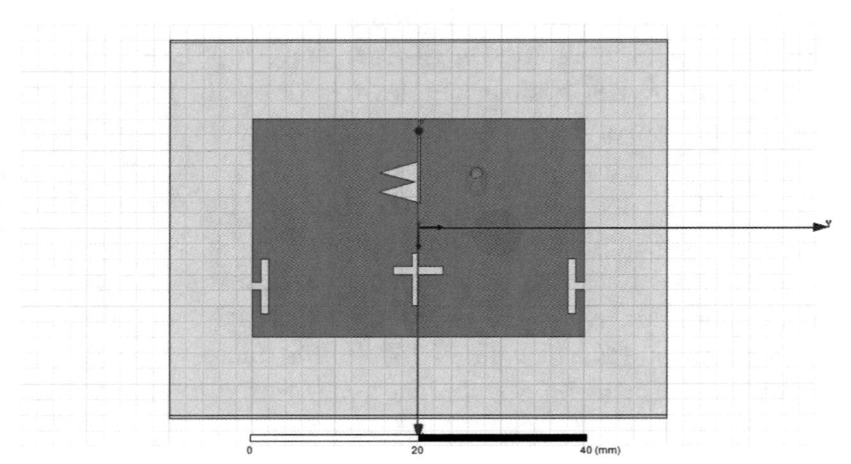

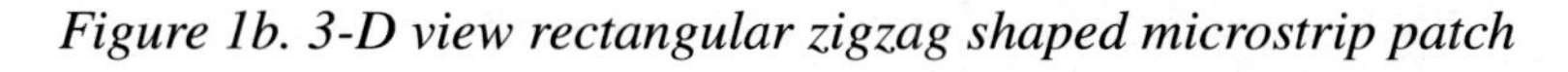

Figure 1b. 3-D view rectangular zigzag shaped microstrip patch

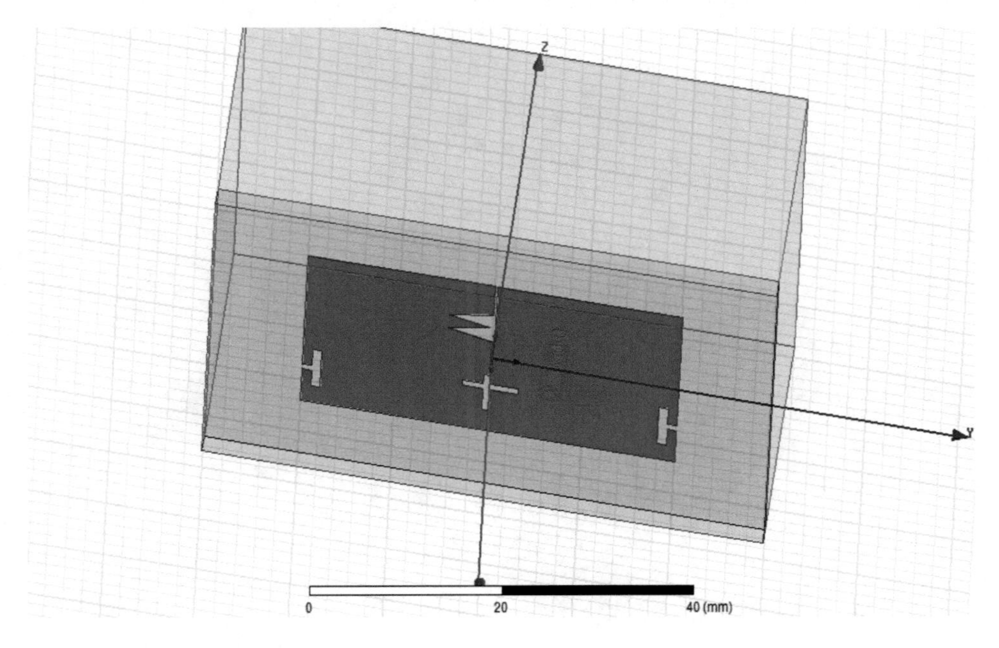

simulation software (HFSS). Figure 1(a) shows the 2-D diagram of rectangular shape of proposed antenna.

The square patch is easily designed widely used easily to analyze and easy to manufacture. To design square patch following method are used

Step-1 Width (w)

$$W = \frac{C}{2 * f_0 \sqrt{\left(\varepsilon_r + 1\right) / 2}} \tag{1}$$

Step-2 Effective Dielectric Constant (ε_{reff})

$$\varepsilon_{reff} = \frac{\varepsilon_r + 1}{2} + \frac{\varepsilon_r - 1}{2} \sqrt{1 + 12 * \left(\frac{h}{2}\right)} \tag{2}$$

Step-3 Effective length (L_{eff})

Figure 1c. Ground view of rectangular zigzag shaped microstrip patch

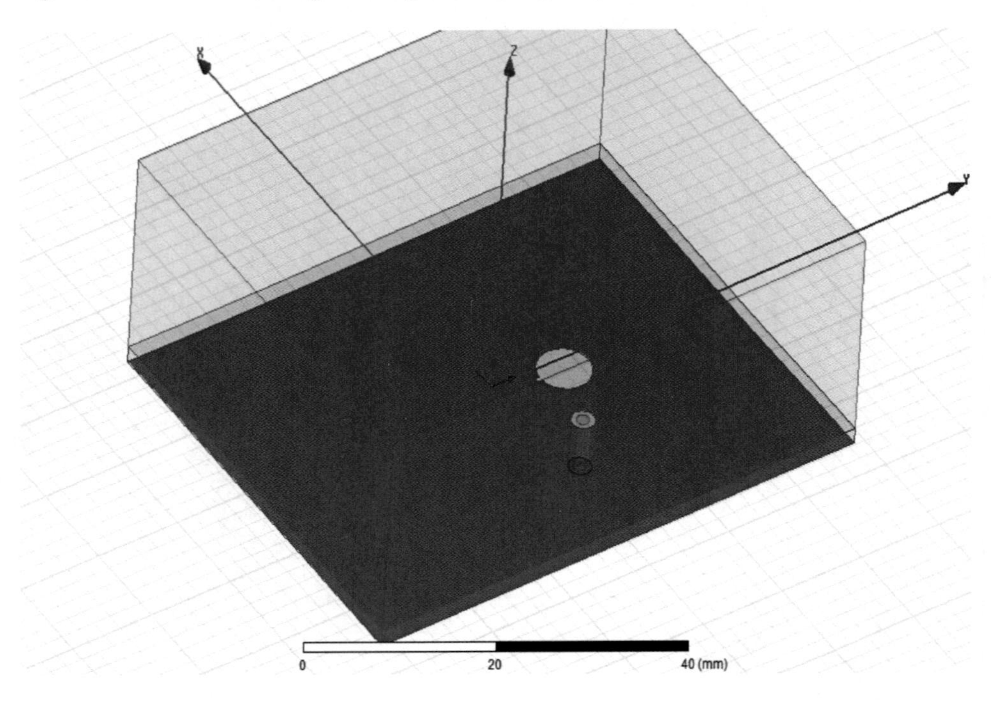

$$L_{eff} = \frac{C}{2 * f_0 * \sqrt{\varepsilon_{eff}}} \tag{3}$$

Step-4 Length extension (ΔL)

$$\Delta L = 0.412 * h \frac{\left(\varepsilon_{eff} + 0.3\right)\left(\frac{w}{h} + 0.264\right)}{\left(\varepsilon_{eff} - 0.264\right)\left(\frac{w}{h} + 0.8\right)} \tag{4}$$

Step-5 Actual length of patch (L)

$$L = L_{eff} - 2\Delta \tag{5}$$

Figure 2. Adaptive mesh refinement of antenna

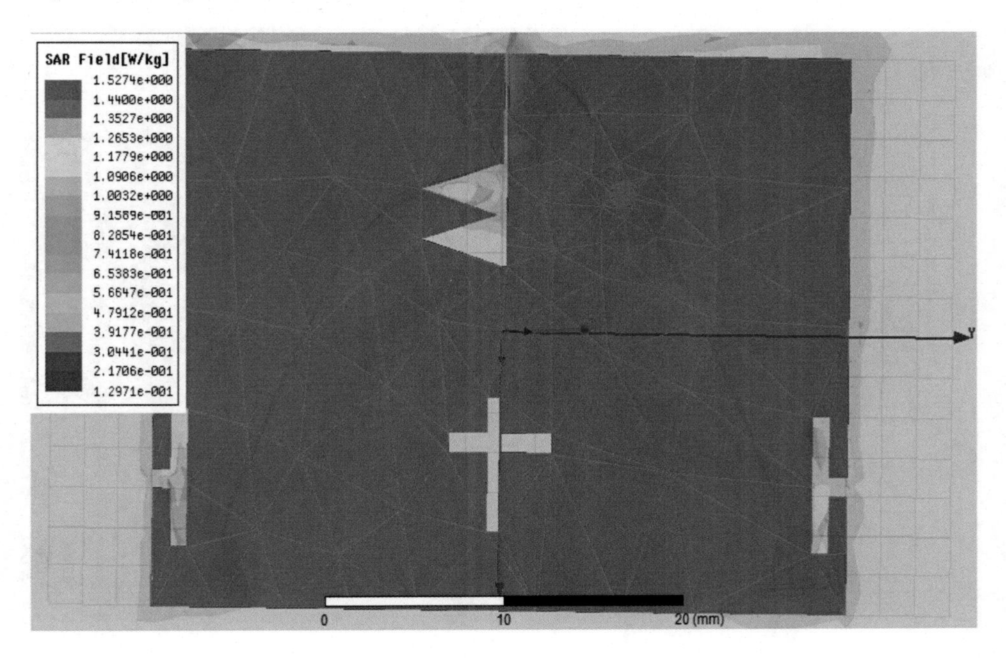

Table 1. Dimension of antenna design

S. No.	Parameter	Dimension (mm.)
1	Substrate	48x60x1.6
2	Ground	48x60
3	Patch	28x40
4	Ground Slot (Radius)	3
5	Distance of Ground Slot for Centre of the modal	9.5mm from Y axis
6	Feed type - Probe	Inner radius 0.7
		Outer radius 1.3
		X(-7), Y(+7)

The diagram 1 (b) shows the 3-D diagram of proposed antenna that is designed. In this figure clearly show that the entire two T-section. One Zig-Zag with cross cut also added the circular defected ground structure (DGS).

The figure 1 (c) shows the back sides view of proposed antenna. This figure shows the patch of the antenna is clearly view in the backside view of the proposed antenna.

Figure 3. Working models of HFSS with proposed antenna

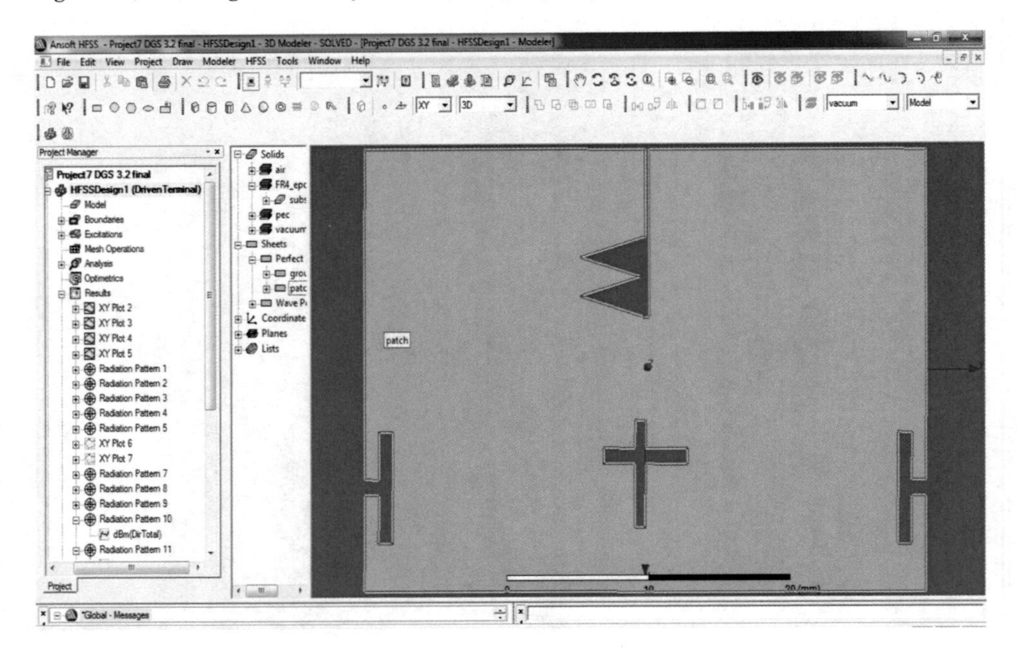

In figure 2 the adaptive mesh refinement of the proposed antenna is to be shown to generate the correct solution. The dimension of Proposed antenna is described in table 1.

Figure 3 describe the HFSS window with proposed design that is designed, validate and optimized on high frequency software simulator. The design of proposed antenna has been completed using Ansoft HFSS-13 software. This simulation gives the satisfactory result for hardware implementation. This design can used for the 1 to 5 GHz frequency range and give the different result for return loss VSWR, active Z parameter, Impedance BW (MHz), antenna gain (in decibel) and other results. In the proposed work we have also simulate the three different results which are electromagnetic wave direction figure, adaptive mesh refinement of antenna and SAR field diagram. All these three results are very important if we design any micro strip antenna because at last finally when we talk about hardware implementation of our proposed design SAR values play an important role, SAR directly affected to the human body so we try to minimize as low as values of SAR.

RESULTS AND DISCUSSION

Return Loss (S-11)

$$S(11) = -10 \log_{10} \frac{\mathrm{Pr}}{Pi}$$

The prototype design has stimulated by using software HFSS13. All the comparable results of both the software's are achieved by simulation and approximation for proposed design.

VSWR

Figure 4. Simulated return losses (S-11) of the proposed antenna

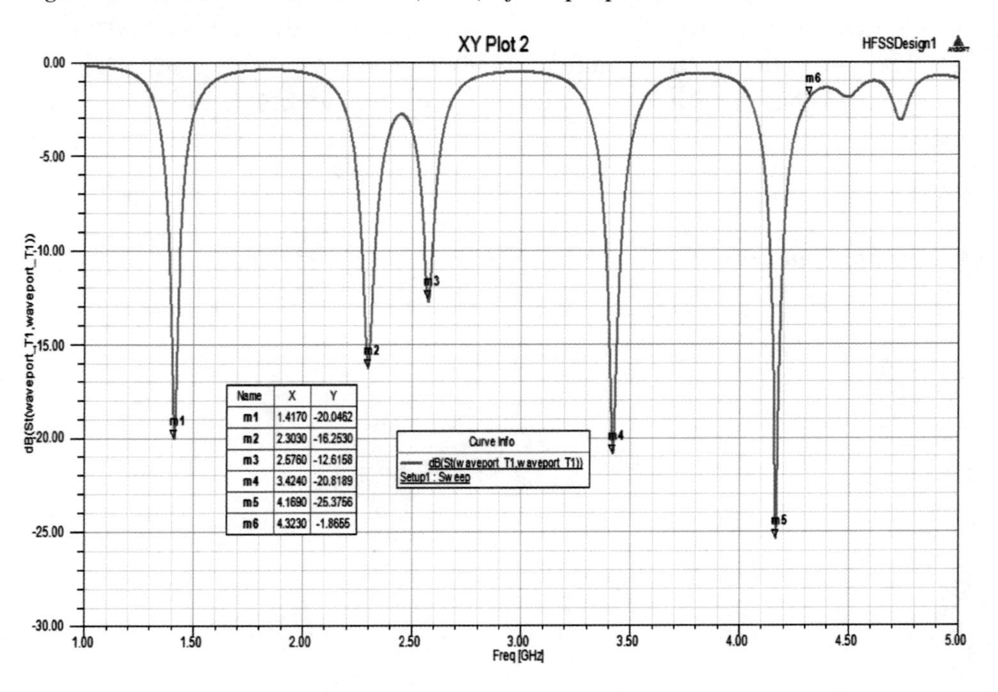

The VSWR is can be determine by using formula which is

$$VSWR = \frac{1+\Gamma}{1-\Gamma} \tag{6}$$

RADIATION PATTEN OF PROPOSED ANTENNA

The radiation pattern explain about the changes of radiated power at arrival angles.

The radiation pattern for 1.42 GHz frequency range is shown in Figure 5 (a). For the resultant output frequency of 1.42 GHz. The phase value is 90 degree and gain in dBm is near to 20dBm (decibel-milliwatts). Futher it convrted into a dBi values that is 10dBi. The Decibel isotropic (dBi) is the forward gain of an antenna compared with the hypothetical isotropic antenna, which uniformly distributes energy in all directions.

Figure 5 (b) shows the radiation pattern for 2.305 GHz frequency range. At this resultant output frequency 2.305 GHz, the phase value is 90 degree and gain in dBm is near to 29dBm (decibel-milliwatts). Futher it is convrted into a 9 dBi values i.e. in dBi. The dBi (decibel isotropic), the forward gain of an antenna compared with the hypothetical isotropic antenna, which uniformly distributes energy in all directions.

Figure 5a. Radiation pattern for 1.42GHz frequency in HFSS

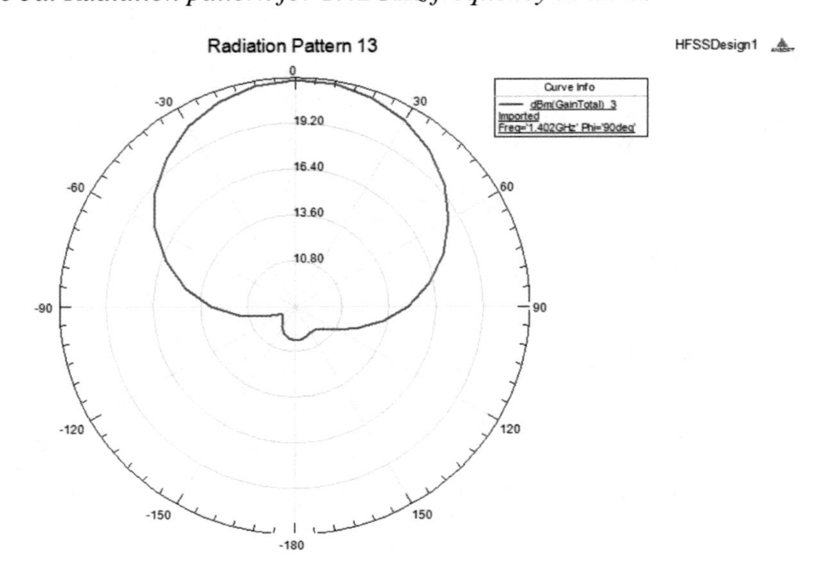

Figure 5b. Radiation pattern for 2.305 GHz frequencies in HFSS

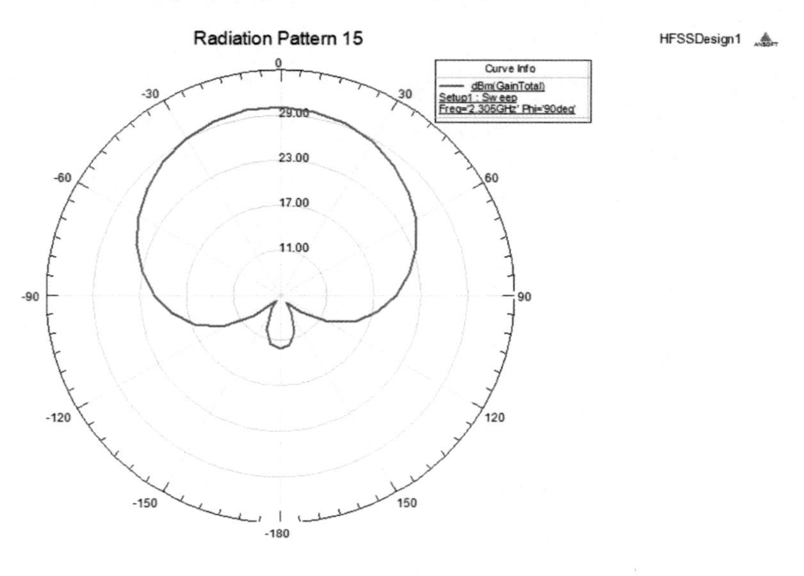

Figure 5c. Radiation pattern for 4.16GHz frequencies in HFSS

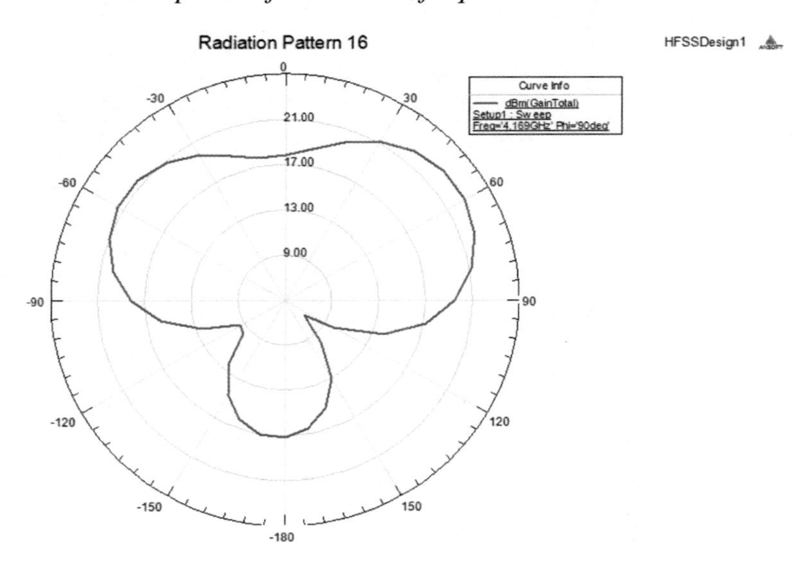

Figure 5 (c) shows the radiation pattern for 4.16 GHz frequency range. At this resultant output frequency 4.169 GHz, phase value is 90 degree and gain in dBm is near to 17 dBm (decibel-milliwatts). Futher it convrted into dBi i.e. 12 dBi. The Decibel isotropic is the forward gain of an antenna compared

Figure 5d. Radiation pattern for 3.41 GHz frequencies in HFSS

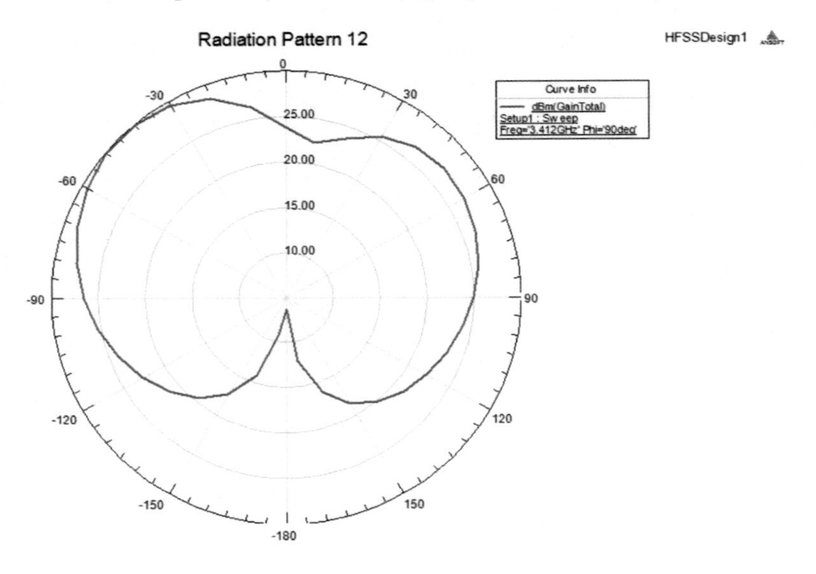

Figure 5e. Radiation Pattern for2.57 GHz frequencies in HFSS-13

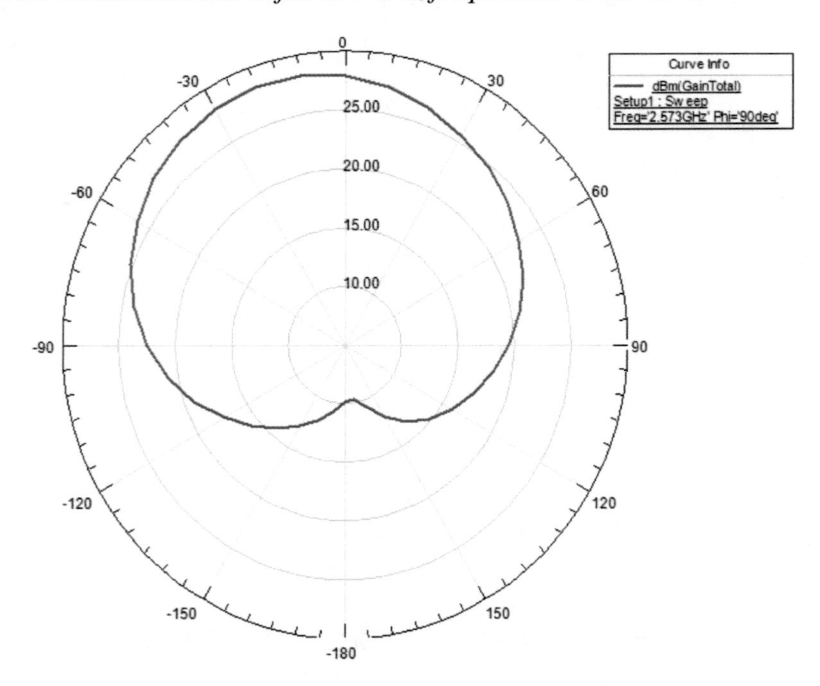

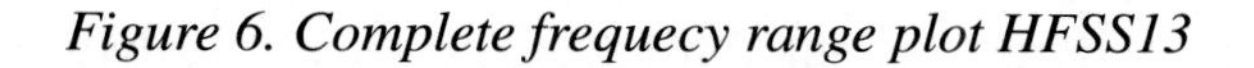

Figure 6. Complete frequecy range plot HFSS13

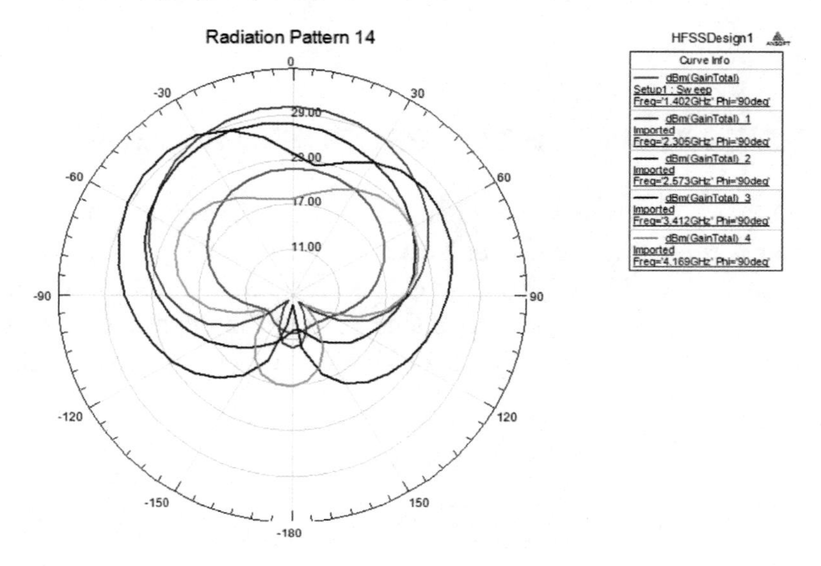

with the hypothetical isotropic antenna, which uniformly distributes energy in all directions.

Figure 5 (d) shows the radiation pattern for 3.41 GHz frequency range. In the resultant output frequency 3.41GHz, phase value is 90 degree and gain in dBm is near to 24dBm(decibel-milliwatts). Futher it convrted into a 6dBi values that is dBi decibel isotropic dBi (decibel isotropic) – the forward gain of an antenna compared with the hypothetical isotropic antenna, which uniformly distributes energy in all directions.

In the next figure we are showing the next frequecy radation patteren for 2.57GHz range. Below Figure 5 (e) shows the radiation pattern for 2.57 GHz frequency range. In the resultant output phase value 90 degree and gain in dBm is near to 26 dBm (decibel-milliwatts). Futher it convrted into a 3 dBi values that is dBi decibel isotropic dBi (decibel isotropic), the forward gain of an antenna compared with the hypothetical isotropic antenna, which uniformly distributes energy in all directions.

Figure 5 (e) Radiation pattern for2.57 GHz frequencies in HFSS-13In Figure 6 shows the final outcome of the all five freqencies in a single Ration pattern plot, that is shown below.

The radiation pattern for different frequencies1.4GHz, 2.3 GHz, 2.5 GHz, 3.42 GHz and 4.16 GHz frequency range are shown in figure from 5(a) to 5(b). In the resultant output frequency is phase value 90 degree and gain in dBm of different frequenies is 24 dBm, 20dBm, 29dBm, 26dBm and 17dBm respectively.

ACTIVE Z-PARAMETER OF PROPOSED ANTENNA

In this, Figure 7 shows active Z parameter of proposed antenna. Active Z parameter is rejonaing parameter by default HFSS 13 software take 50 ohm impedance for the calcualtion of active Z parameter. Antenna tuner or matching circuit is used to match the impedance of both antenna and transmission or feed line . The impedance plot for single geometry antenna is shown in

Figure 7. Active Z-parameter in HFSS13

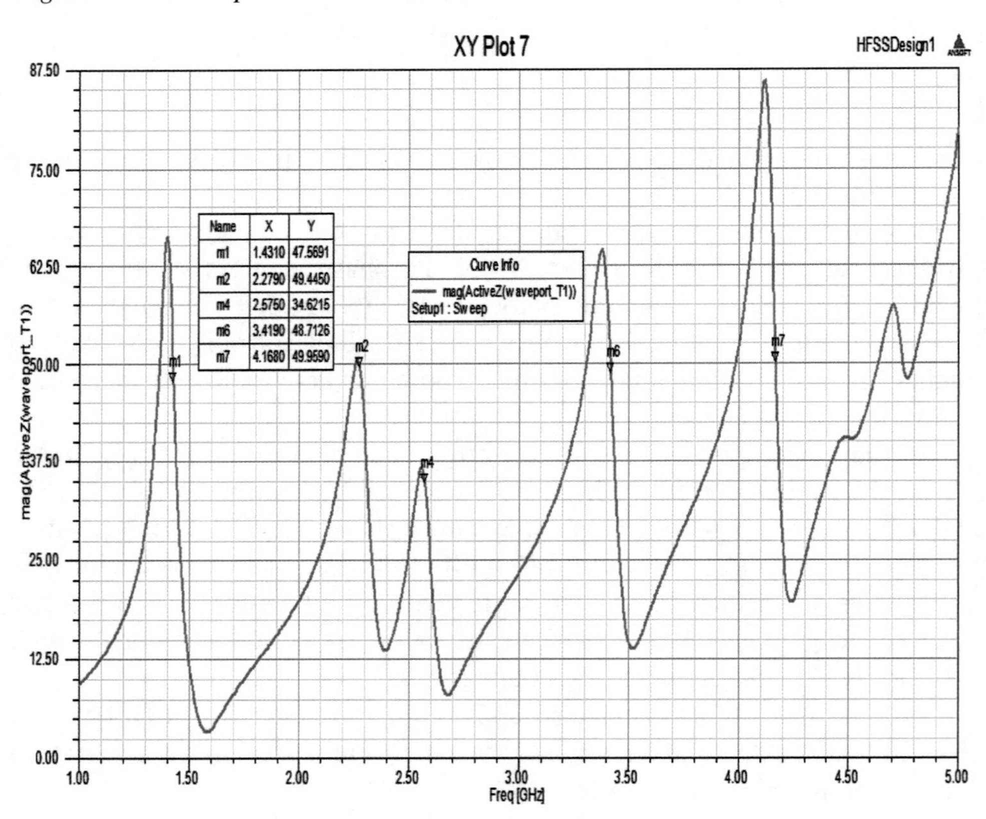

Figure 7. In this figure the five frequecy are shown at the frequency point

Table 2. Shows the active Z parameter

Sr. No.	Frequency Point	Default R (Ohm)	Frequency (GHz)	Active Z impedance (Ohm)
1	m1	50	1.4	47.56
2	m2	50	2.3	49.44
3	m4	50	2.57	34.62
4	m6	50	3.41	48.71
5	m7	50	4.16	49.95

of m1,m2,m3,m4 and m5. When we calculate the HFSS active Z parameter 50 ohm defacult resistancce is used in HFSS 13. Table 2 shows the active Z parameter specified values.

Figure 8. Proposed antenna Poynting vector (power flow) plot and electric field

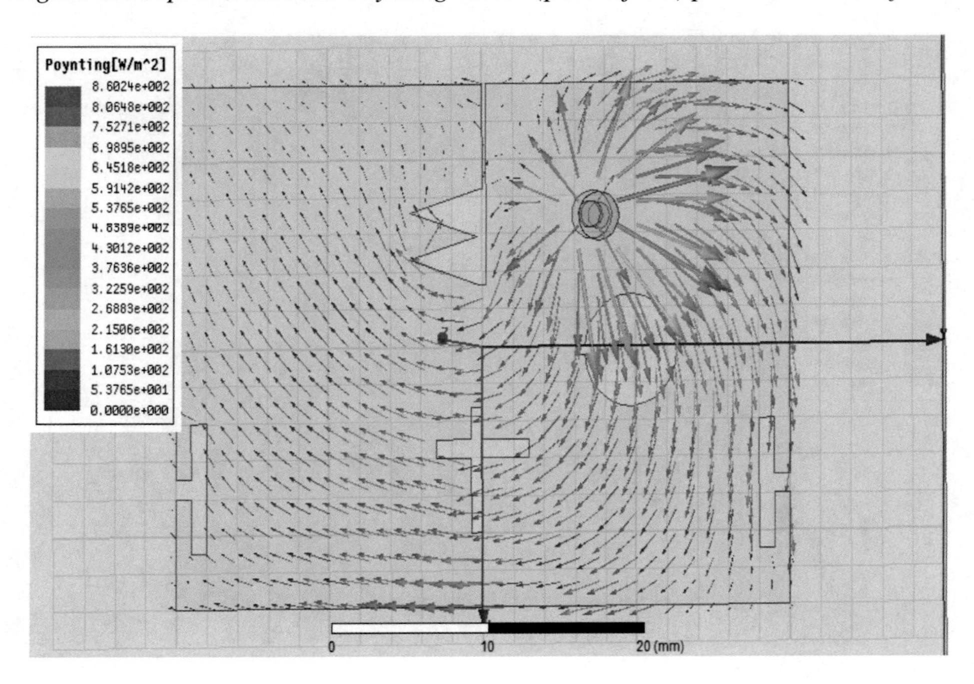

POYNTING VECTOR ((POWER FLOW) PLOT AND ELECTRIC FIELD PATTERN

Poynting vector is the vector product (S = E x H) of the E vector and the H vector. The Poynting vector represents the directional energy flux density (the rate of energy transfer per unit area) of an electromagnetic field. The SI unit of the Poynting vector is the watt per square metre (W/m^2). The proposed antenna Poynting vector (power flow) plot and electric field is shown in Figure 8.

SPECIFIC ABSORPTION RATE RESULTANT OF PROPOSED WORK

Specific absorption rate (SAR) is defined as the power absorbed per mass of tissue and has units of watts per kilogram (W/kg). Figure 9 (a) Shows the SAR result of 3.42 GHz frequency point. This figure shows that the

Figure 9a. SAR result of 3.42 GHz frequency

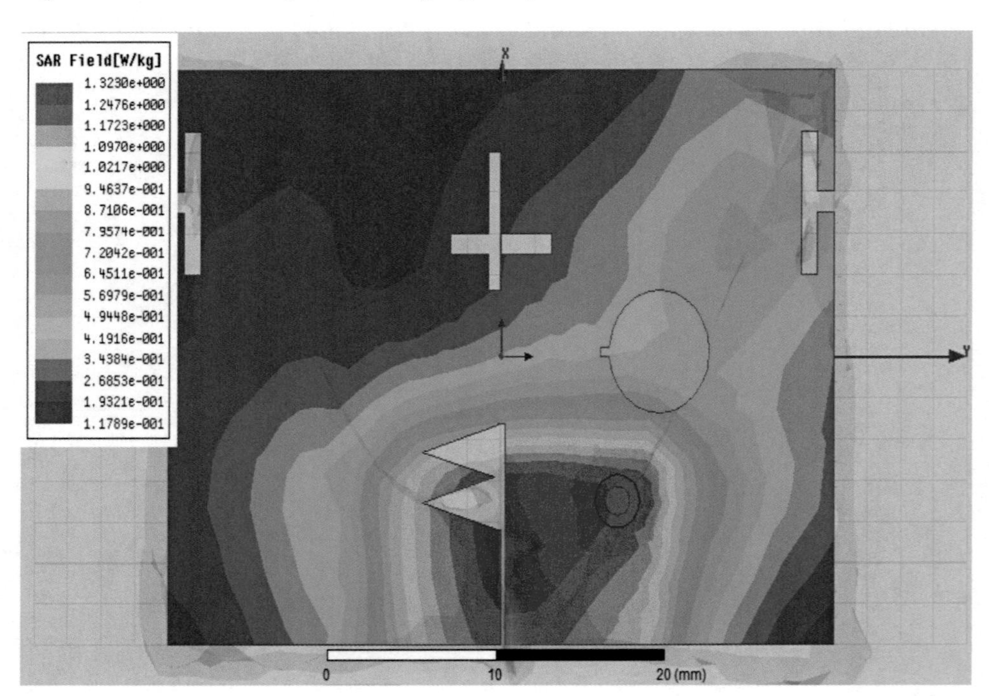

Figure 9b. Shows the SAR result of 2.57 GHz frequency

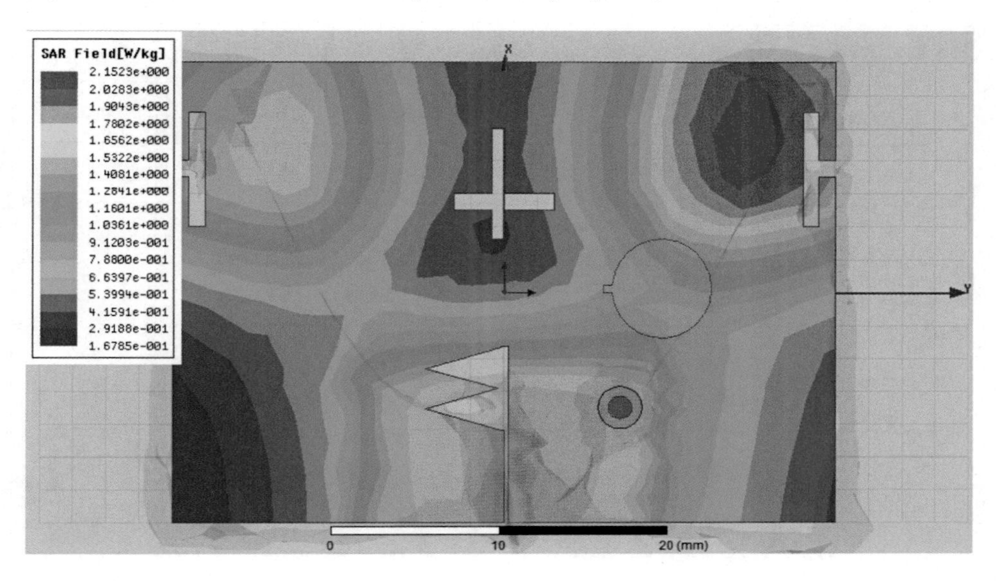

Figure 9c. Shows the SAR result of 1.41 GHz frequency

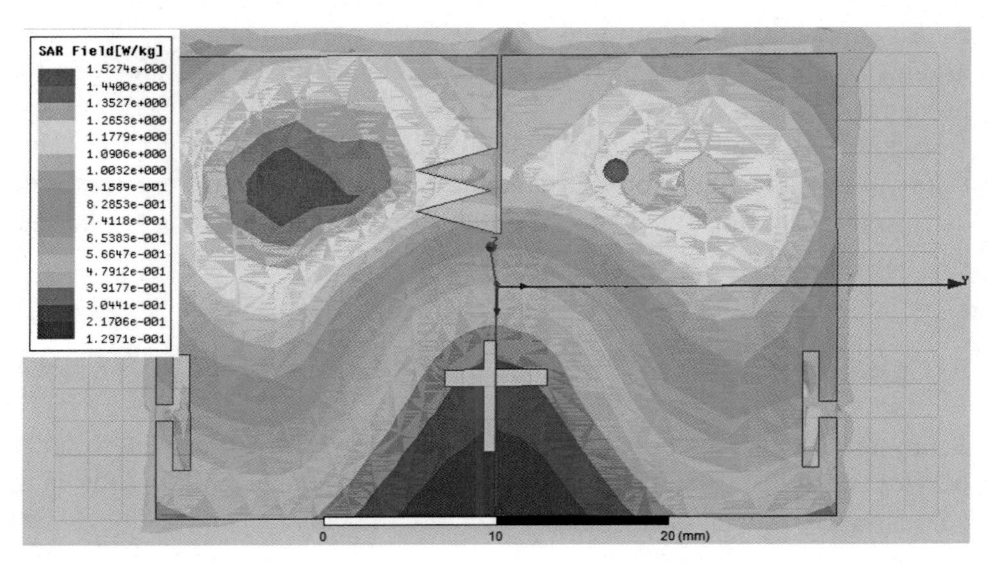

Table 3. Shows the comparison of proposed work with different antenna

Antenna	Structure	Frequency (GHz)	Impedance BW (MHz)	Gain (dBi)
(Liu, Wu, and Dai, 2017)	Monopole antenna with DGS	2.4 3.5 5.2	5.98 .0.92 4.13	2.46 2.45 3
(Pei et. al, 2011)	Tri Band Antenna with DGS	2.61 3.5 5.4	0.3 1.05 0.96	1.85 2.19 2.57
(Dong, Toyao, and Itoh, 2011)	Antenna with Split ring resonators	2.31 2.83	1.38 3.29	3.13 3.85
(Reddy and Vakula, 2014)	Slit loaded antenna with circular DGS	2.45 3.5 5.28	6.5 4.18 4.97	4.72 6.2 3.8
(Chakraborty et. al, 2017)	High Performance DGS Integrated Compact Antenna for 2.4/5.2/5.8 GHz WLAN Band	2.4 5.2 5.8	1.0 8.0	2.02 5.9 4.14
Proposed Method	Zigzag Shaped Microstrip Patch Antenna with Cross Cut-set	**1.4 2.3 2.5 3.42 4.16**	5.1 5.7 3.6 5.9 5.9	4.85 5.6 3 5.78 5.45

proposed antenna SAR plot contain very less area around 70% to 80% area of our proposed antenna under the safe zone for human and natural things.

Figure 9 (b) Shows the SAR result of 2.57 GHz frequency point. This figure shows that the proposed antenna SAR plot contain very less area around 70% to 80% area of our proposed antenna under the safe zone for human and natural things.

Figure 9 (c) Shows the SAR result of 2.57 GHz frequency point. This figure shows that the proposed antenna SAR plot contain very less area 70% to 80% area of our proposed antenna under the safe zone for human and natural things.

9. COMPARISON OF RESULT WITH DIFFERENT ANTENNA

Table 3 represents the comparison of proposed work with respect to other pervious designed IEEE Tranistion and Letters antenna. In this table, the main comparison parameters are frequecy (GHz), impedence band width (MHz) and gain (dBi) of different antenna. There are different antenna with different structure are shown in column 2 and in column 3, it shows the frequency range in giga hetrz. In this table the different types antenna are shown with its geometry and also with DGS.

CONCLUSION

Multiband zigzag shaped micro-strip patch antenna with Cross Cut-set patch antenna with defected ground structure is designed, simulated, and tested for wireless applications. Antenna structure with and without defected ground plane is presented and compared. The difference between the bandwidths at different frequency bands is small, which is an attractive feature. The gain values at the operating frequencies of novel structure are also measured. The measured and simulated radiation patterns are in good agreement. This study elucidates the tradeoff between compactness through DGS with gain. The proposed antenna is applicable for wireless communication systems with enhanced bandwidth especially working in Wi-max and wireless application. These results shows antennas could be developed for possible applications in several wireless systems like WLAN and Wi-MAX if properly scale to the allowed frequency bands as well as research for the Military Field. This proposed antenna is very useful in point to point communication area. This designed antenna is for high power, radio frequency (R.F.) efficient radio equipped to transmit over entire frequency between the ranges of 1.4 to 4.68 GHz frequency also which is also compatible for OFDM & military purposes.

REFERENCES

Al-Hasan, M. J., Denidni, T. A., & Sebak, A. R. (2012). EBG Dielectric-Resonator Antenna with Reduced Back Radiation for Millimeter-Wave Applications. *IEEE International Symposium on Antennas and Propagation Society*, 1(2), 702-702. 10.1109/APS.2012.6348942

Aras, M. S. M., Rahim, M. K. A., Asrokin, A., & Aziz, M. A. (2008). Dielectric resonator antenna (DRA) for wireless application. *2008 IEEE International RF and Microwave Conference*, 454-458. 10.1109/RFM.2008.4897461

Bhardwaj, A., & Kumar, V. D. (2012). Optical dielectric resonator antenna. *9th International Conference on Communications (COMM)*, 159-162.

Chakraborty, M., Chakraborty, S., Reddy, P. S., & Samanta, S. (2017). High performance DGS integrated compact antenna for 2.4/5.2/5.8 GHz WLAN band. *Wuxiandian Gongcheng*, *26*(1), 71–77.

Coulibaly, Y., & Denidni, T. A. (2009). Gain and bandwidth improvement of an aperture coupled cylindrical dielectric resonator antenna using an EBG structure. *IEEE International Symposium on Antennas and Propagation Society*, 1-4.

Dong, Y., Toyao, H., & Itoh, T. (2011). Design and characterization of miniaturized patch antennas loaded with complementary split-ring resonators. *IEEE Transactions on Antennas and Propagation*, *60*(2), 772–785. doi:10.1109/TAP.2011.2173120

El-Deen, E., Zainud-Deen, S. H., Sharshar, H. A., & Binyamin, M. A. (2006, July). The effect of the ground plane shape on the characteristics of rectangular dielectric resonator antennas. In *2006 IEEE Antennas and Propagation Society International Symposium* (pp. 3013-3016). IEEE. 10.1109/APS.2006.1711242

Ge, Y., & Esselle, K. P. (2009). A dielectric resonator antenna for UWB applications. *2009 IEEE Antennas and Propagation Society International Symposium*, 1-4. 10.1109/APS.2009.5172064

Kaur, J., Khanna, R., & Kartikeyan, M. (2014). Novel dual-band multistrip monopole antenna with defected ground structure for WLAN/IMT/BLUETOOTH/WIMAX applications. *International Journal of Microwave and Wireless Technologies*, *6*(1), 93–100. doi:10.1017/S1759078713000858

Khalily, M., Kamarudin, M. R., Mokayef, M., & Jamaluddin, M. H. (2014). Omnidirectional circularly polarized dielectric resonator antenna for 5.2-GHz WLAN applications. *IEEE Antennas and Wireless Propagation Letters*, *13*, 443–446. doi:10.1109/LAWP.2014.2309657

Li, T. W., & Sun, J. S. (2005). Wideband dielectric resonator antenna with parasitic strip. *IEEE/ACES International Conference on Wireless Communications and Applied Computational Electromagnetics*, 376-379.

Liu, W. C., Wu, C. M., & Dai, Y. (2011). Design of triple-frequency microstrip-fed monopole antenna using defected ground structure. *IEEE Transactions on Antennas and Propagation*, *59*(7), 2457–2463. doi:10.1109/TAP.2011.2152315

Luk, K. M. (2003). *KW" Leung, Dielectric Resonator Antennas. Research Studies*. Hertfordshire, UK: Press Ltd.

Mittal, N., Khanna, R. G., & Kaur, J. G. (2015). *Design Analysis and Fabrication of Microstrip Patch Antennas for Various Applications Using Electromagnetic Bandgap and Defected Ground Structures* (Doctoral dissertation).

Mu'ath, J., Denidni, T. A., & Sebak, A. R. (2013). Millimeter-wave EBG-based aperture-coupled dielectric resonator antenna. *IEEE Transactions on Antennas and Propagation*, *61*(8), 4354–4357. doi:10.1109/TAP.2013.2262667

Pei, J., Wang, A. G., Gao, S., & Leng, W. (2011). Miniaturized triple-band antenna with a defected ground plane for WLAN/WiMAX applications. *IEEE Antennas and Wireless Propagation Letters*, *10*, 298–301. doi:10.1109/LAWP.2011.2140090

Reddy, B. S., & Vakula, D. (2014). Compact zigzag-shaped-slit microstrip antenna with circular defected ground structure for wireless applications. *IEEE Antennas and Wireless Propagation Letters*, *14*, 678–681. doi:10.1109/LAWP.2014.2376984

Ryu, K. S., & Kishk, A. A. (2010). Ultra wideband dielectric resonator antenna with broadside patterns mounted on a vertical ground plane edge. *IEEE Transactions on Antennas and Propagation*, *58*(4), 1047–1053. doi:10.1109/TAP.2010.2041160

Singh, A. K., Kabeer, R. A., Shukla, M., Singh, V. K., & Ali, Z. (2013). Performance Analysis of First Iteration Koch Curve Fractal Log Periodic Antenna of Varying Angles. *Open Engineering, 3*(1), 51-57.

Singh, V. K., Dhupkariya, S., & Bangari, N. (2017). Wearable Ultra Wide Dual Band Flexible Textile Antenna for WiMax/WLAN Application. *International Journal of Wireless Personal Communications, 95*(2), 1075–1086.

Yadav & Singh. (2019). Design of U-shape with DGS circularly polarized wearable antenna on fabric substrate for WLAN and C-Band applications. *Journal of Computational Electronics, 18*, 1-7.

Yang, F., & Rahmat-Samii, Y. (2003). Reflection phase characterizations of the EBG ground plane for low profile wire antenna applications. *IEEE Transactions on Antennas and Propagation*, *51*(10), 2691–2703. doi:10.1109/TAP.2003.817559

Chapter 8
Performance Analysis of Cascade H-Bridge Multilevel Inverter Topology With Filter Circuit and Without Filter Circuit

Nikhil Agrawal
Ujjain Engineering College, India

ABSTRACT

Multilevel inverter is a modified version of inverter. Multilevel inverter recently emerged in the area of high power and medium voltage application. In the last few decades, the great innovation has been done to improve the inverter performance, and it is challenging even today. The multilevel inverter performance is examined by total harmonic distortion and component required. In multilevel, as level increases, the total harmonic distortion value decreases, but the number of components required and driver circuit increases that make the circuit more complex and also the effect on cost. So, the challenge is to balance the bridge between cost and total harmonic distortion. This chapter simulates the various levels of conventional cascade H-bridge inverter and new proposed topology of multilevel inverter with using different modulation techniques and with using filter circuit and without filter circuit.

DOI: 10.4018/978-1-5225-9683-7.ch008

INTRODUCTION

The objective of Inverter is to produce an AC output power from a DC power source at rated voltage and frequency. The DC source can be fuel cell, battery, renewable energy source. The output of ideal inverter should be AC sinusoidal but in practice the output obtain AC non-sinusoidal waveforms that contains large number of harmonics. These harmonics degrade the performance of inverter and also produce power quality issue at consumer end. These harmonics content minimized by modifies the inverter circuit. Inverter circuit basically two level. Multilevel inverter is unfurled version of inverter circuit (Agrawal, Singh, and Bansal, 2017). Multi-level inverter starts basically from three level. Two level inverter produce square wave output i.e. it is much more deviated from sinusoidal waveform and THD value about 48 percent is much higher than IEEE standard limit that produces power quality issue at consumer end. To overcome these problem multilevel inverter used. As level increases the square waveform converts in staircase sine waveform so by which it contains less harmonics and also THD value lowered (Rodriguez, Jih-Sheng, and Zheng, 2002). To minimize THD value further various scheme adopted such as filter circuit and various PWM techniques. But problem arise in increase the level, the component required more and also cost associated with. So obtain less THD value in output different topology invented.

MULTILEVEL INVERTER

The multi-level inverter has invented in 1970-1980s (Baker, 1980). The basic concept of multi-level inverter is to obtain high power and voltage by using a power semiconductor switches. Multilevel inverter produce staircase voltage waveform. To eliminate the harmonics many scheme adopted such as multicarrier modulation techniques and filter circuit. The modulation techniques compared reference wave i.e. sinusoidal wave (fundamental frequency wave) with carrier wave (higher frequency triangular wave) and output pulse is intersection of these signals and it is responsible for switching ON or switching OFF the inverter switching device.

The several advantage over a two- level inverter. The brief introduction of these advantages as follows (Ebrahimi, Babaei and Gharehpetion, 2012; Colak, Kabalci, Bayindir, et al., 2011).

- **Output Waveform Quality:** Multilevel inverter provides less distorted waveform and that reduced the $\frac{dv}{dt}$ stress.
- **Common Mode Voltage:** High common mode voltage is responsible for failure of motor bearings that is connected to multilevel inverter drive circuit. The multilevel inverter produce the low common mode voltage.
- **Less Total Harmonic Distortion**: Multilevel inverter produces low value of harmonics in output waveform.
- **Switching Frequency:** Multilevel inverter have a switching device that will operate at low switching frequency as low switching frequency that means low switching loss so that efficiency is high in case of multilevel inverter.

Multilevel inverter also have some disadvantages compare to two level inverter.

- **Large number of Switching Devices:** As level increases to lowered the THD factor in output voltage waveform the number of power semiconductor switch and their driver circuit require more that will circuit become complex, bulky, difficulty in packaging and also effects on cost.
- **Chances of System Failure**: Due to complexity of circuit, there are more chances of system failure.

Although multi-level inverter have these disadvantage still multilevel inverter is first choice for high power medium voltage application.

The multilevel inverter topology can be classified as-

1. Monolithic inverter
2. Modular inverter

Monolithic Multilevel Inverter

Monolithic multilevel inverter structure based on DC source arrangement to generate staircase output. Monolithic multilevel inverter categories as diode clamped inverter and flying capacitor inverter.

Diode Clamped Inverter

Diode clamped inverter firstly presented in 1981 by A. Nabae, I. Takahashi and H. Akagi. In this topology diode is used is used as clamping device to clamp the dc voltage to achieve staircase output waveform. This topology uses fundamental switching frequency i.e. switching losses less so that efficiency is high. All phase share a common dc bus, which minimize the capacitor requirements (Takahashi and Akagi, 1981). The main drawback of this topology is that maximum output voltage is half of the supply voltage.

Flying Capacitor Inverter

Flying capacitor is presented in 1992 by meynard and fochin. The structure of this topology is similar to previous topology, the only difference is that clamping capacitor used instead of clamping diode (Meynard and Foch, 1992).

Modular Multilevel Inverter

Modular refer as consisting of units or separate parts that can be joined together. The one unit is the full wave two level inverter which is look like alphabetical letter 'H'. Modular MLI cascade the converter cell to reach high power high voltage output so modular MLI is also knows as Cascade H-Bridge MLI (Sniha and Lipo, 1996). The cascade H- Bridge MLI introduced in 1975 by Lai and Peng.

Different MLI Topology Based on Cascade H- Bridge Multilevel Inverter

Topology 1: Topology 1 is conventional Cascade H- Bridge multilevel inverter (CHBMLI).it is very popular topology. It is the series connection of single phase H- bridge unit. It requires 2N-2 power switches for N level inverter.

Topology 2: In this topology one cell of H-Bridge connected to complementary switches and voltage sources connected in alternative manner for level addition and subtraction (Sahoo, Ramulu, Prakash and Deeksha, 2012). This topology needed N+3 power switches for N level inverter.

Figure 1. (a) Basic Unit of Topology 1; (b) 7 level structure of Topology 1

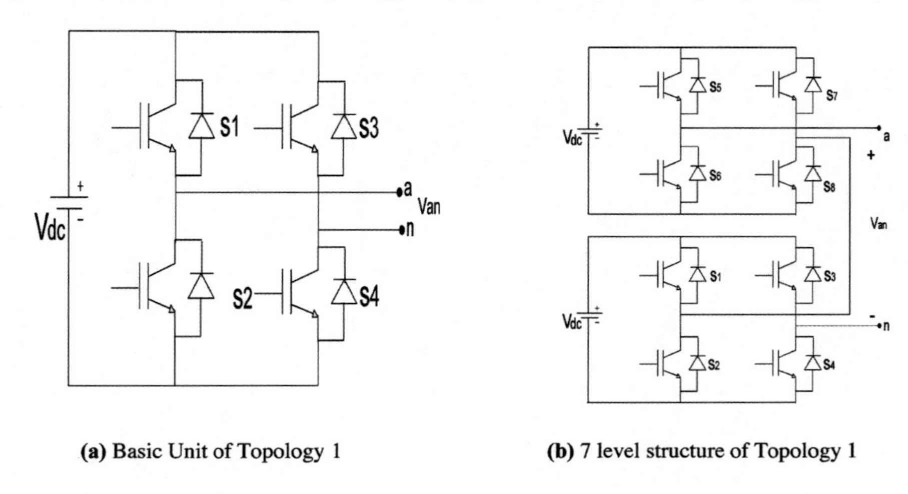

(a) Basic Unit of Topology 1

(b) 7 level structure of Topology 1

Figure 2. (a) Basic unit MLI structure of Topology 2; (b) 7 level structure of topology 2

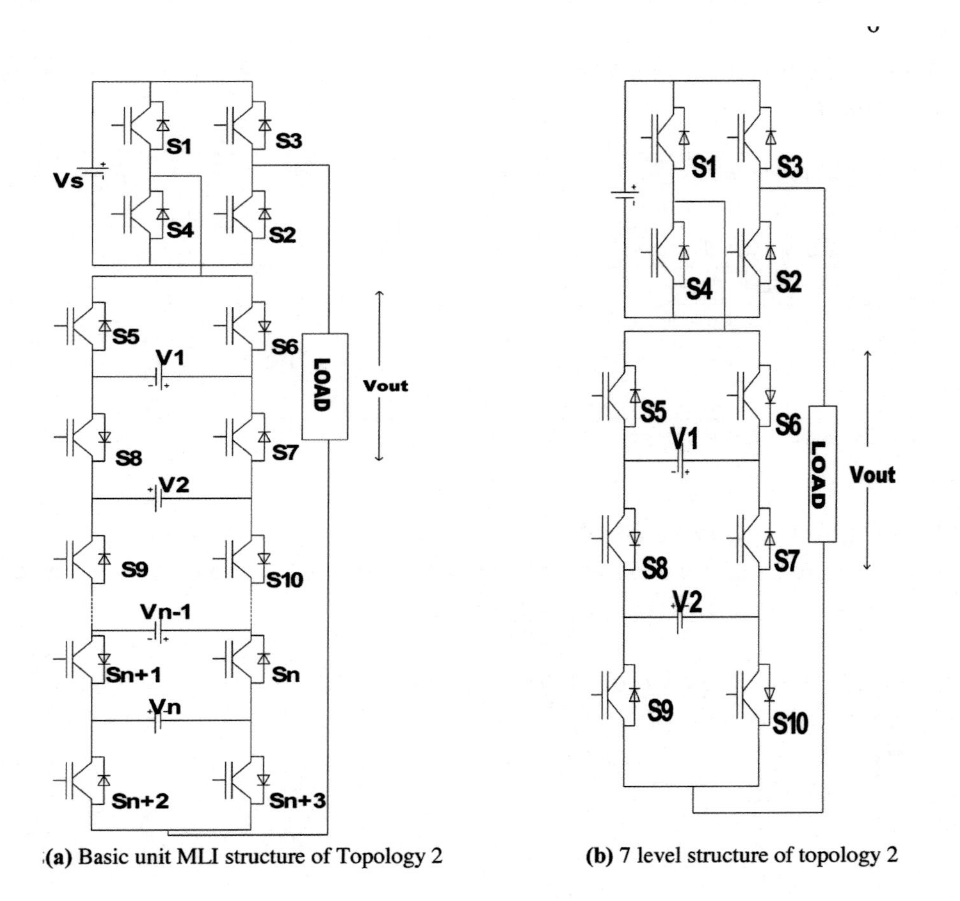

(a) Basic unit MLI structure of Topology 2

(b) 7 level structure of topology 2

Figure 3. 7 level structure of MLI of Topology 3

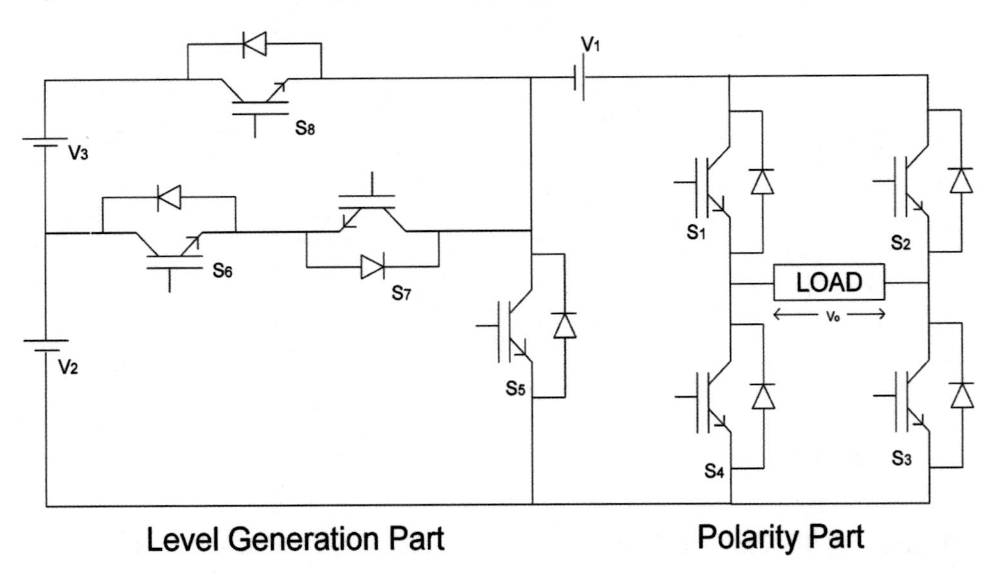

Table 1. Comparison of power component in different multilevel inverter topologies for N level structure

MLI Component	Topology 1	Topology 2	Topology 3
Main Switch Device	2(K-1)	K+3	K+1
Clamping Capacitor	0	0	0
Clamping Diode	0	0	0
DC Sprit Capacitor	0	0	0
DC Source	(K-1)/2	(K-1)/2	(K-1)/2
Total	5(K-1)/2	(3K+5)/2	(3K+1)/2

Where K= Number of level

Topology 3: Proposed topology consist of two part first is level generation part and second is polarity part (Samsami, 2017). This topology needed N+1 power switches for N level inverter.

For 7 level topology 1 require total 15 components, topology 2 require total 13 components and topology 3 require only 11 components. Topology 1 is conventional MLI topology. Topology 2 and topology 3 is derived modify cascade H-bridge MLI topology.

MODULATION TECHNIQUES

Modulation techniques used to reduce the harmonics in the output waveform (Holmes and McGrath, 2001). Without modulation technique a lot of calculation are required to generate the pulse pattern for switches so applying modulation techniques calculation can be avoided. In this article the carrier based modulation technique and hybrid modulation techniques are applied.

Phase Disposition Pulse Width Modulation Techniques (PD PWM)

PD modulation techniques is carrier based modulation technique. For K level configuration requires (K-1) triangular carrier waves. In the Phase Disposition method all (K-1) carrier have same frequency and amplitude with same phase (Rushirai and Kapil, 2016). In the carrier-based implementation, at every instant of time the modulating signal wave are compared with the carrier wave and depending on comparator function, the switching pulses are generated. For seven level multilevel inverter there are 6 triangular wave, out of 6, 3 triangular wave above the zero reference wave and other 3 below the zero reference wave and reference set zero is placed to middle of carrier set as shown in fig.

Figure 4. PDPWM technique carrier arrangement

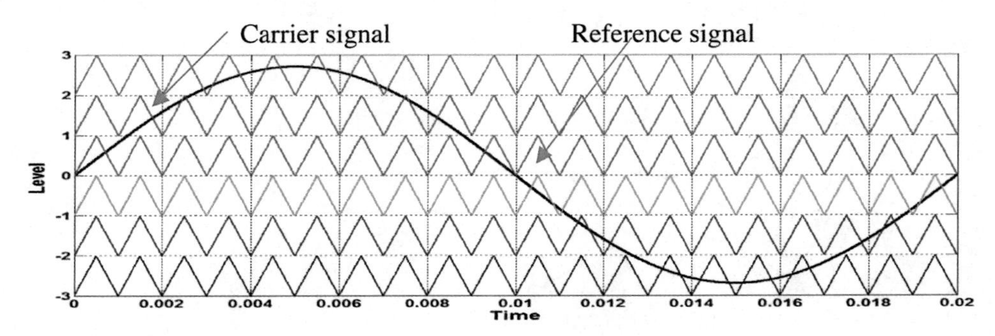

Phase Opposition Disposition Pulse Width Modulation Techniques (POD PWM)

POD- PWM modulation technique is carrier based modulation technique (Najafi, Yatim and Samosir, 2008). In this techniques the carrier signals are in same phase to each other above and below the zero reference axis with same amplitude and frequency and carrier signals below the zero reference axis is 180° out of phase to above the zero- axis carrier waveform (Wu, n.d.) as shown in fig.

Figure 5. PODPWM technique carrier arrangement

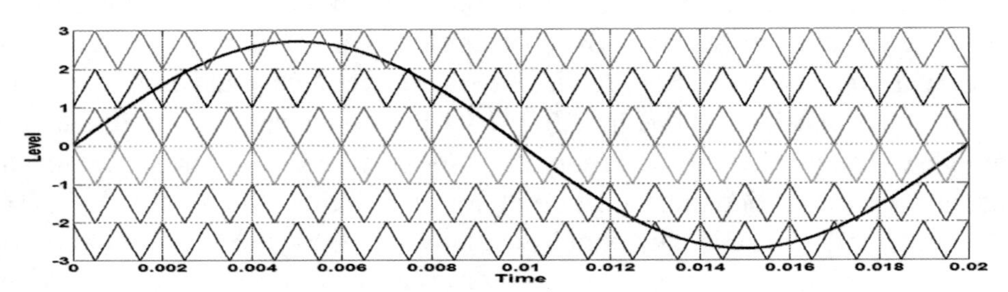

Alternate Phase Opposition Disposition Pulse Width Modulation Techniques (APOD PWM)

APOD modulation technique also carrier based modulation technique. In Alternate phase opposition disposition PWM technique all (K-1) carrier signals

Figure 6. APOD PWM technique carrier arrangement

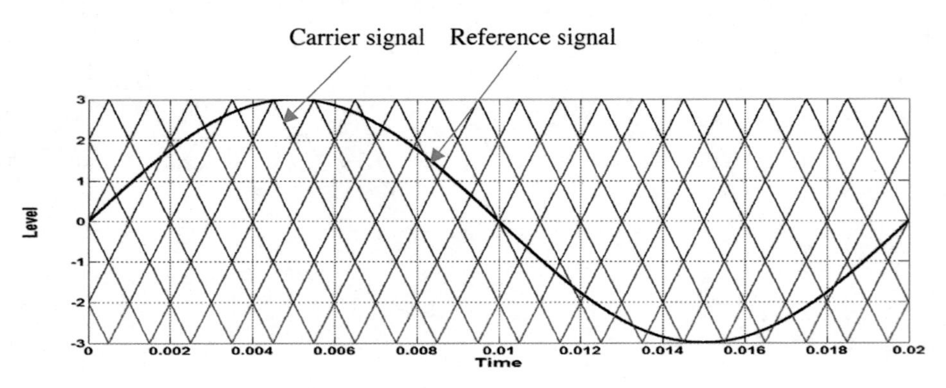

are phase-shifted 180° alternatively from each other but all carrier signals have same amplitude and frequency i.e. all carrier have same frequency and amplitude and every carrier is in out of phase with its neighbor carrier by 180° (Babaei, 2010; Najafi and Yatim, 2012).

Inverted Sine Carrier Pulse Width Modulation Techniques (ISC PWM)

Inverted sine carrier pulse width modulation technique is Hybrid modulation Technique. ISC PWM technique replace the level shifted based Carrier waveform by inverted sine wave (Babaei, 2008). The inverted sine carrier PWM technique uses the sine wave as a reference signal and an inverted half sine wave with carrier signal

Figure 7. ISC PWM technique carrier arrangement

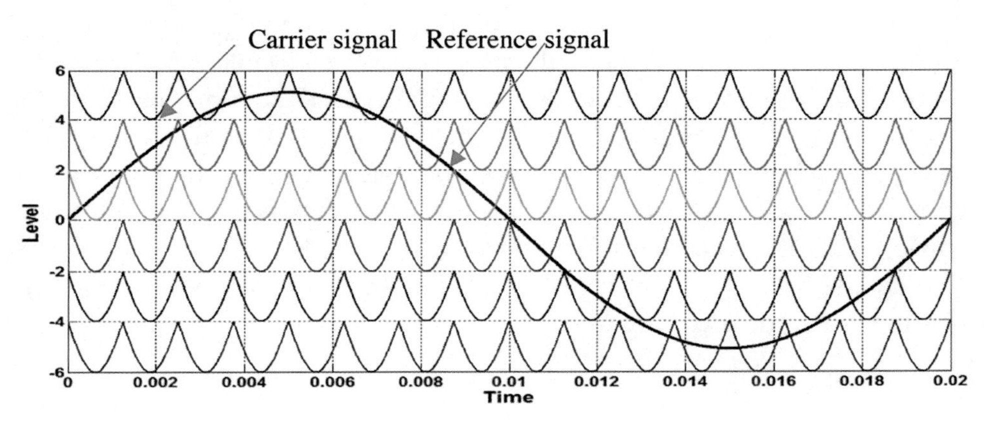

Inverted Sine Carrier with Variable Frequency Pulse Width Modulation Techniques (ISCVF PWM)

Inverted sine carrier with variable frequency modulation technique is classification of hybrid modulation techniques (Gui-Jia, 2005). In this PWM technique sine wave as a reference signal and carrier signals are inverted sine signal with variable frequency as shown in fig. This VFISCPWM techniques provide enhanced fundamental voltage and lower THD.

Figure 8. VFISC PWM technique carrier arrangement

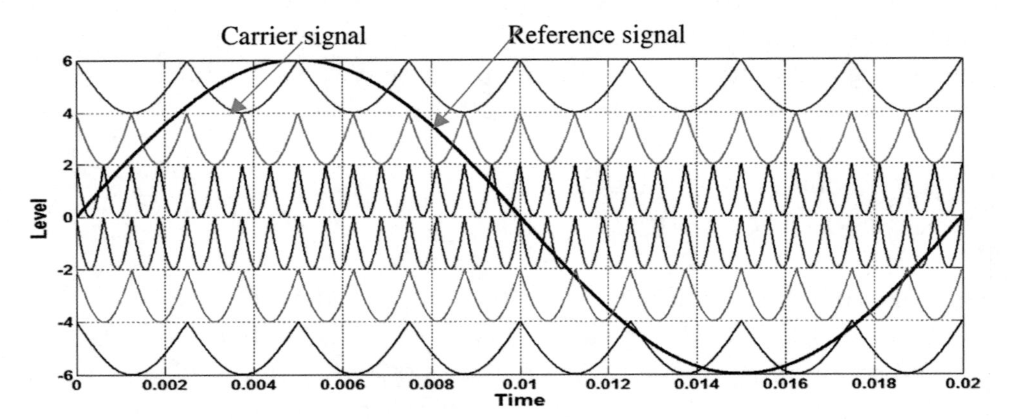

SIMULATION RESULT

This chapter presents the comparison of THD for single phase seven level different cascade H- Bridge based multilevel inverter topologies. These MLI topologies implemented in MATLAM/SIMULINK 2010b. The switch is selected as an IGBT bidirectional switch. The carrier wave frequency taken as 2 KHz and reference wave frequency is 50Hz. Each DC source is of 10V and load is resistive of 5 ohm and filter component inductance of 10mH and capacitor of 1microfarad is taken and simulated for 4 cycle i.e. 0.08 seconds. The detailed simulation result is shown in table 2 for without filter circuit and in table 3 for filter circuit.

Figure 9. (a) THD of topology 1 MLI with PD PWM; (b) THD of topology 1MLI with POD PWM;(c) THD of topology 2 MLI with PODPWM; (d) THD of topology 2 MLI with ISCPWM;(e) THD of topology 2 MLI with VFISCPWM; (f) THD of topology 3 MLI with APOD PWM;Total Harmonic Distortion of Proposed MLI topologies without filter circuit

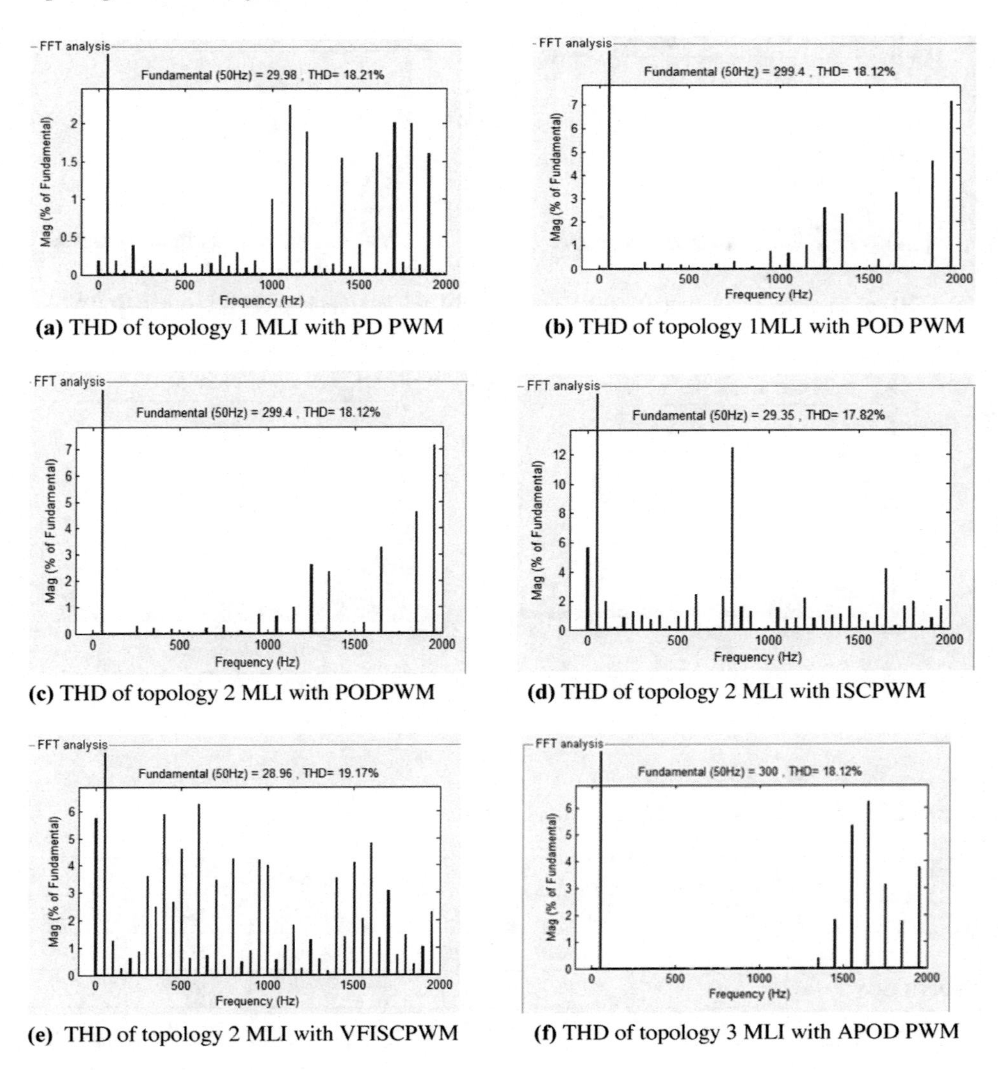

(a) THD of topology 1 MLI with PD PWM

(b) THD of topology 1MLI with POD PWM

(c) THD of topology 2 MLI with PODPWM

(d) THD of topology 2 MLI with ISCPWM

(e) THD of topology 2 MLI with VFISCPWM

(f) THD of topology 3 MLI with APOD PWM

Figure 10. (a) THD of topology 1MLI with PD PWM; (b) THD of topology 1 MLI with POD PWM;(c) THD of topology 2 MLI with APOD PWM; (d) THD of topology 3 MLI with PD PWM;Total Harmonic Distortion of proposed MLI topologies with Filter circuit

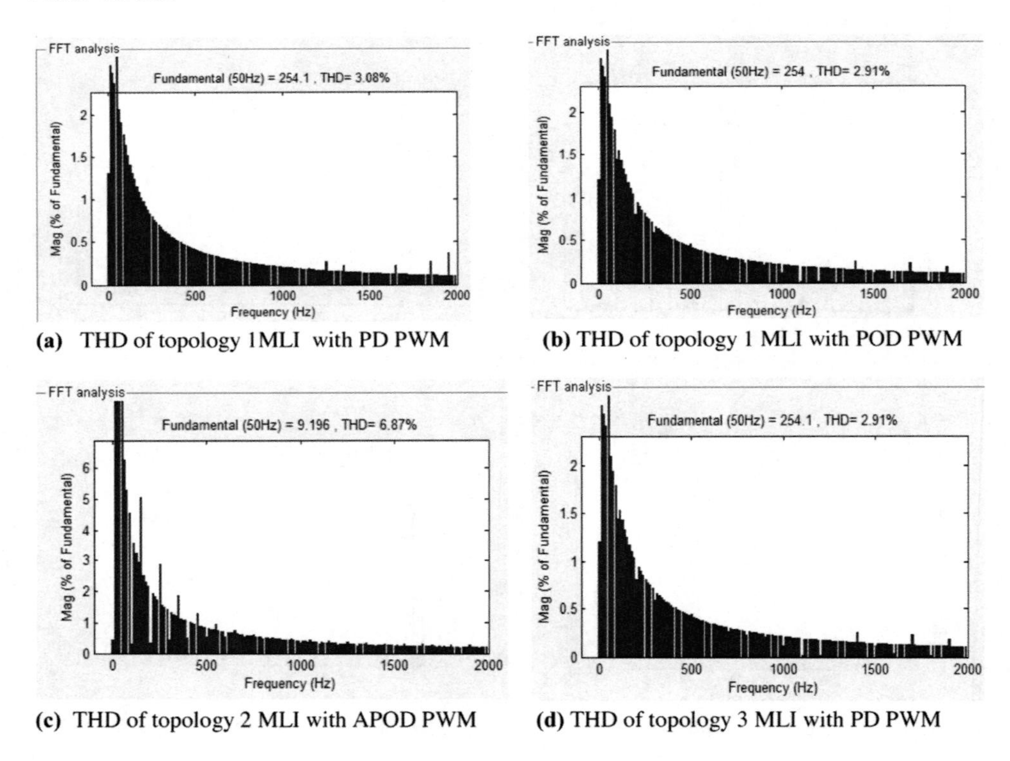

(a) THD of topology 1MLI with PD PWM

(b) THD of topology 1 MLI with POD PWM

(c) THD of topology 2 MLI with APOD PWM

(d) THD of topology 3 MLI with PD PWM

Table 2. % THD result in MLI topologies without LC filter circuit

MLI Topology	Modulation Index	Modulation Techniques				
		PD	POD	APOD	ISC	ISCVF
TOPOLOGY 1	0.8	24.36	24.22	24.18	23.57	24.40
	0.9	22.42	22.15	22.08	20.35	20.12
	1.0	18.19	17.95	18.25	17.81	19.13
TOPOLOGY 2	0.8	24.33	24.00	24.12	23.60	24.39
	0.9	22.38	22.05	21.92	20.30	20.02
	1.0	18.21	18.12	18.34	17.82	19.17
TOPOLOGY 3	0.8	24.36	24.23	24.18	23.57	24.40
	0.9	22.42	22.15	22.08	20.35	20.12
	1.0	18.19	17.95	18.12	17.81	19.13

Table 3. % THD result in MLI topologies with LC filter circuit

MLI Topology	Modulation Index	Modulation Techniques				
		PD	POD	APOD	ISC	ISCVF
TOPOLOGY 1	0.8	3.15	3.04	3.15	4.61	4.30
	0.9	3.14	3.03	3.13	3.92	3.73
	1.0	3.08	2.91	3.08	3.90	3.53
TOPOLOGY 2	0.8	8.94	8.49	8.44	7.57	9.54
	0.9	8.76	8.09	8.12	7.24	8.85
	1.0	8.34	7.70	6.87	6.59	8.14
TOPOLOGY 3	0.8	3.04	3.14	3.14	4.61	4.30
	0.9	3.02	3.14	3.12	3.92	3.73
	1.0	2.91	3.07	3.08	3.90	3.53

CONCLUSION

Multilevel inverter topology performance judged by component require and total harmonic distortion factor. Simulation results of proposed Multilevel Inverter topologies as topology 1, topology 2 and topology 3 for single phase 7- Level with carrier disposition modulation techniques and hybrid modulation technique with different modulation index is calculated in MATLAB/Simulink. Proposed MLI topologies required total components data is shown in Table1 and a comparison made among proposed multilevel inverter topologies without LC filter circuit is shown in Table 2 and with LC filter circuit is shown in Table 3. MLI Topology 3 requires less component as comparisons to topologies 1 and topology 2.The output of MLI is passed through LC filter and assess the output waveform near to sinusoidal wave with THD value lowered. Above these result it is conclude that MLI topology 3 with filter circuit have total harmonic distortion factor (THD) is under the maximum permissible limit according to IEEE standard.

REFERENCES

Agrawal, Singh, & Bansal. (2017). A Multilevel Inverter Topology using Reverse-Connected Voltage Sources. *International Conference on Energy, Communication, Data Analytics and soft Computing (ICECDS 2017)*, 1290-1295.

Babaei, E. (2008). A cascade multilevel converter topology with reduced number of switches. *IEEE Transactions on Power Electronics*, *23*(6), 2657–2664. doi:10.1109/TPEL.2008.2005192

Babaei, E. (2010). Optimal topologies for cascade sub-multilevel Converters. *J. Power Electron.*, *10*(3), 251–261. doi:10.6113/JPE.2010.10.3.251

Baker. (1980). *High-voltage converter circuits*. U.S. Patent Number 4,203,151.

Colak, I., Kabalci, E., & Bayindir, R. (2011). Review of multilevel voltagesource inverter topologies and control schemes. *Energy Conversion and Management*, *52*(2), 1114–1128. doi:10.1016/j.enconman.2010.09.006

Ebrahimi, J., Babaei, E., & Gharehpetion, G. B. (2012). A new Multilevel converter topology with reduced number of Power electronic components. *IEEE Transactions on Power Electronics*, *59*(2), 655–667.

Holmes & McGrath. (2001). Opportunities for harmonic cancellation with carrier-based PWM for two-level and multilevel cascaded inverters. *IEEE Trans. Ind. Applicate.*, *37*, 574–582.

Meynard, T. A., & Foch, H. (1992). Multilevel choppers for high Voltage applications. *Proc. European Conf. Power Electronics and applications*, 45-50.

Najafi, E., & Abdul, H. M. Y. (2012, November). Design and Implementation of a New Multilevel Inverter Topology. *IEEE Transactions on Industrial Electronics*, *59*(11), 4148–4154. doi:10.1109/TIE.2011.2176691

Najafi, E., Yatim, A. H. M., & Samosir, A. S. (2008). A newtopology-reversing voltage (RV) for multi-levelinverters. *2nd International conference on power andenergy (PECon 08)*, 604-608.

Rodriguez, J., Jih-Sheng, L., & Zheng, F. (2002). Multilevel inverters: A survey of topologies, controls, and applications. *IEEE Transactions on Industrial Electronics*, *49*(4), 724–738. doi:10.1109/TIE.2002.801052

Rushiraj & Kapil. (2016). Analysis of Different Modulation techniques for multilevel inverters. *ISTIEEE International conference on power electronics, intelligent control and energy system.*

Sahoo, Ramulu, Prakash, & Deeksha. (2012). Performance Analysis and Simulation of five level and seven level single phase multilevel inverters. *Third International conference on Sustainable energy and intelligent system (SEISCON 2012).*

Samsami, H., Taheri, A., & Samanbakhsh, R. (2017). New bidirectional multilevel inverter topology with staircase cascading for symmetric and asymmetric structures. *IET Power Electronics*, *10*(11), 1315–1323. doi:10.1049/iet-pel.2016.0956

Sinha, G., & Lipo, T. A. (1996). A four level Rectifier- inverter System for drive Applications. *Conf. Rec. 31St IEEE IAS Annual Meeting*, *2*, 980–987.

Su, G.-J. (2005). S. gui- jia, multilevel dc-link inverter. *IEEE Transactions on Industry Applications*, *41*(3), 848–845. doi:10.1109/TIA.2005.847306

Takahashi & Akagi. (1981). A New neutral point Clamped PWM Inverter. *IEEE Trans. Ind. Applicant.*, *1*(17), 518-523.

Chapter 9
Rhombus-Shaped Cross-Slot and Notched Loaded Microstrip Patch Antenna

Gaurav Varma
Bundelkhand Institute of Engineering and Technology Jhansi, India

Rishabh Kumar Baudh
https://orcid.org/0000-0003-1026-2373
Bundelkhand Institute of Engineering and Technology Jhansi, India

ABSTRACT

The aim is to design a Rhombus microstrip patch antenna. The antenna operates at FL=1.447 GHz to FH=2.382 GHz frequency for wireless local area network (WLAN). This antenna operates at f=1.914 GHz resonant frequency. In microstrip patch antenna, many types of shapes like circular, triangular, rectangular, square, ring shape, etc. are used, but in this design a rectangular shape is used. In proposed antenna, the accuracy and efficiency are increased. Integral equation three-dimensional (3D) software (IE3D) is used for the optimize of the rhombus cross-slotted antenna design. The IE3D uses a full wave method of moment simulator. This antenna fabricated on FR4 glass epoxy double-sided copper dielectric material with relative permittivity of ∈ =4.4, thickness h= 1.60mm, and loss tangent is 0.013.

DOI: 10.4018/978-1-5225-9683-7.ch009

INTRODUCTION

Microstrip rhombus patch antennas contain many properties, includes, light weight, low profile, low cost, less volume and easy to install on rigid surface due to these properties it is easy to fabricate. In microstrip patch antenna, substrate is used as an insulator. The substrate is having fixed dielectric constant. The patch and ground made by copper (Balanis, 1997 & James and Hall, 1989 & Mailloux, McIlvenna, and Kemweis, 1981). In microstrip patch antenna by using substrate having low dielectric constant, thick substrate, using slots (U and L shape), using multilayer and using parasitic patch above the radiating patch to improve bandwidth (Pues and Van de Capelle, 1984 & Chadha and Kumar, 2012 & Ansoft Corp). The microstrip antenna has been considered to be the most inventive field in the engineering of antenna having the properties like low material cost and the simulation and fabrication process is easy. So an idea to use dielectric material as a radiator was perceived. These antennas have use in space applications, government and commercial applications (Singh, Dhupkariya, Bangari, 2017 & Singh et. al, 2013 & Yadav and Singh, 2019). They include radiating patch of metallic material on substrate with ground structure on its back.

In the presented prototype design, It has been used a rectangular microstrip rhombus shape patch antenna which contain a parasitic patch, of length [Lp] and width [Wp] of the patch. All these were calculated by predefine equations. In this design the length and the width of the patch is same and then cut the slots so the one slot of patch at which connect the feed by probe feeding and another patch work as parasitic patch. Application bands are defined by using the software known by IE3D with a S11 of -17.81dB. In this antennadesigntheIE3Dsoftwaresimulatedthefrequencyband from 1.447GHz to 2.382GHz, the WLAN application required this band. The simulation result, we can see that the microstrip patch antenna with parasitic patch find better results. This design achieves percentage bandwidth about 48%.

ANTENNA DESIGN AND MATERIAL

A dielectric constant of this proposed antenna is 4.4, height of the dielectric substrate h=1.6mm and loss tangent tan δ=.012 are selected for the designing of this antenna. The designing frequency (*fr*) is 2.4 GHz.

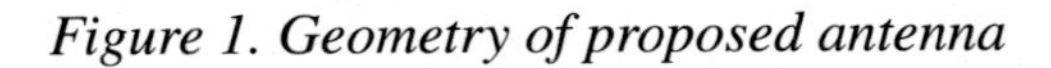

Figure 1. Geometry of proposed antenna

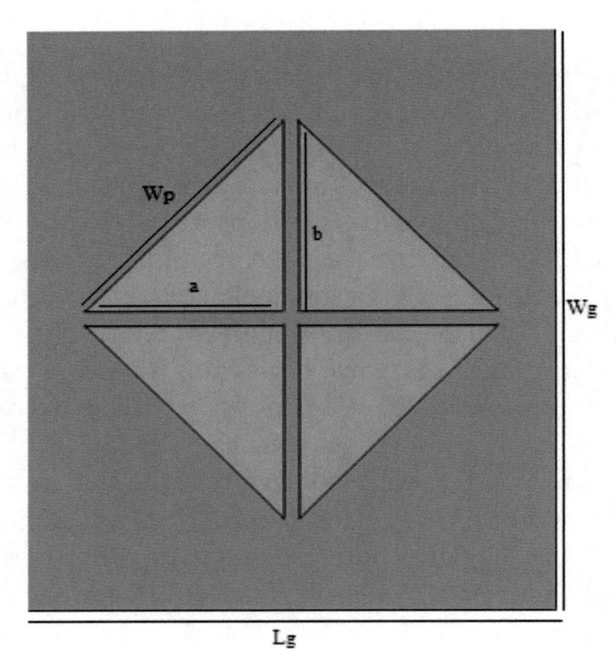

For designing a rectangular Microstrip patch antenna the width and length are calculated as below (Chadha and Kumar, 2012). The width of the rectangular patch can be calculated

$$W = \frac{0.5C}{f}\sqrt{\frac{2}{\varepsilon_r + 1}}$$

Where C = Light's speed ($3\text{x}10^{11}$ mm/s), f = Antenna design frequency, ε_r = Dielectric constant of substrate

The effective dielectric constant ε_{re} is calculated as

$$\varepsilon_{re} = \frac{\varepsilon_r + 1}{2} + \frac{\varepsilon_r - 1}{2}\left[1 + 12\frac{h}{W}\right]^{-0.5}$$

The extended patch length ΔL is calculated as and actual value of the patch length is calculated as follows

$$L = \frac{c}{2fr_{\sqrt{\varepsilon reff}}} - 2\Delta L$$

$$L = \frac{0.5C}{f\sqrt{\varepsilon_{re}}} - 2\Delta L$$

DESIGN PROCEDURE OF ANTENNA

Proposed figure 1, the geometry consists of the rhombus micro strip patch antenna. This consists of rhombus patch with dimension of length Lp=40mm, width Wp=40mm and rotate 45anglesclockwise.The rhombus patch other side with ground plane with (FR4) Epoxy dielectric substrate with relative permittivity of ∈=4.4, and thickness of this material is 1.6mm.The specifications of all notches and slots are given in the table1 (all measurement in mm).

The measurements of ground plane and patch are started from (0, 0) at center for designing the proposed antenna in IE3D software tool. To achieve proper impedance matching and maximum bandwidth, a 50 Ω co-axial feed has been fed at a point (34.25, 66.15) inside the top of patch.

Table 1. Design specifications of proposed antenna

S.No.	Parameters	Value (mm)
1.	H	1.60
2.	L g	70.00
3.	W g	80.00
4.	L p	40.00
5.	W p	40.00
6.	Slot Q (L, W)	(2,56.5)
7.	Slot R (L, W)	(56.5,2)

RESULTS

The proposed antenna is analyzed through Zeeland IE3D simulator. This antenna covers range of frequency about 1.447 GHz to 2.382 GHz as a lower cutoff to highest cutoff frequency respectively with return loss of -17.81dB. It displays maximum return loss of -17.81dB at 1.914GHz, illustrated in figure 2. According to analysis of return loss graph, it produced the bandwidth about 48%with 1.914GHz as a center frequency. Proposed geometry of antenna is obtained after various modifications such as slotting and notching inside the rectangular patch.

This propose antenna has high antenna efficiency about 91%at 2.1GHz which shown in Figure.6. Smith Chart is a graphical representation of impedance as well as radiation of the proposed antenna. Normalized Pattern of Radiation is simulated to obtain other parameters like gain and half power beam width. Stimulated radiation pattern is displayed in figure 8. 2 D radiation pattern is measured at phi 0° and 90°.

Figure 2. BW impedance of projected antenna

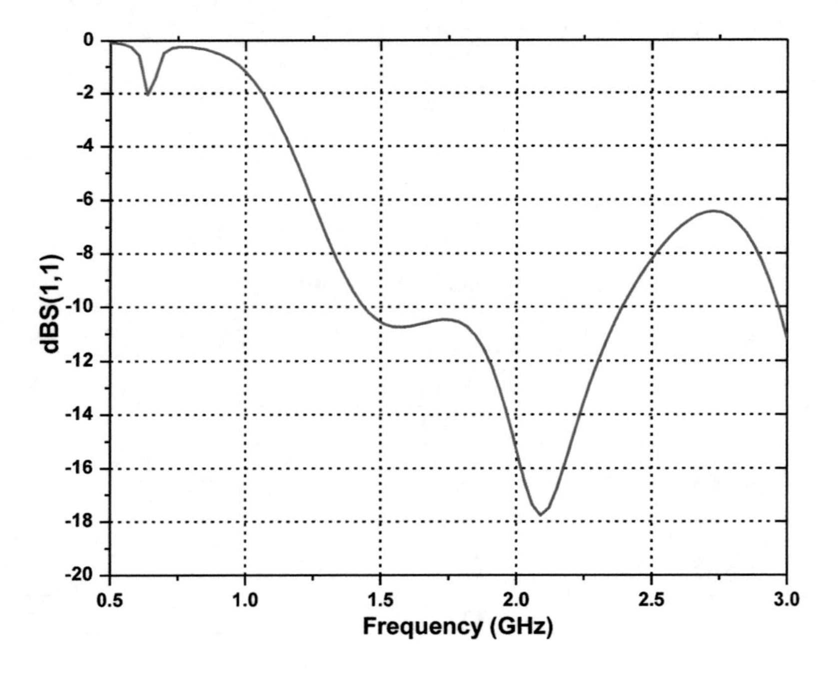

Figure 3. VSWR of proposed antenna

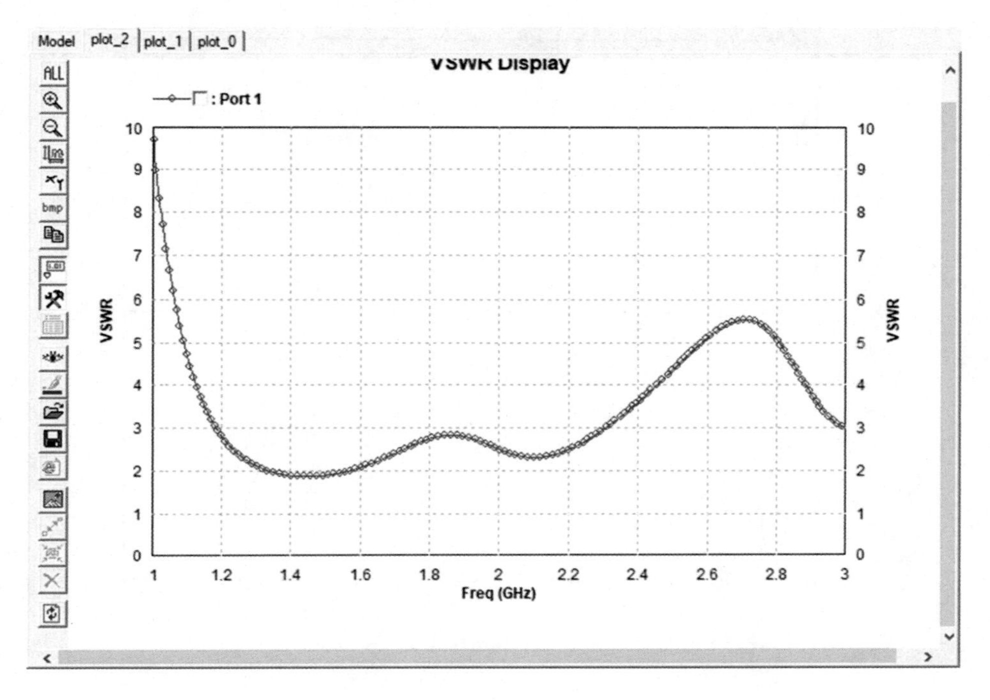

Figure 4. Directivity of proposed antenna

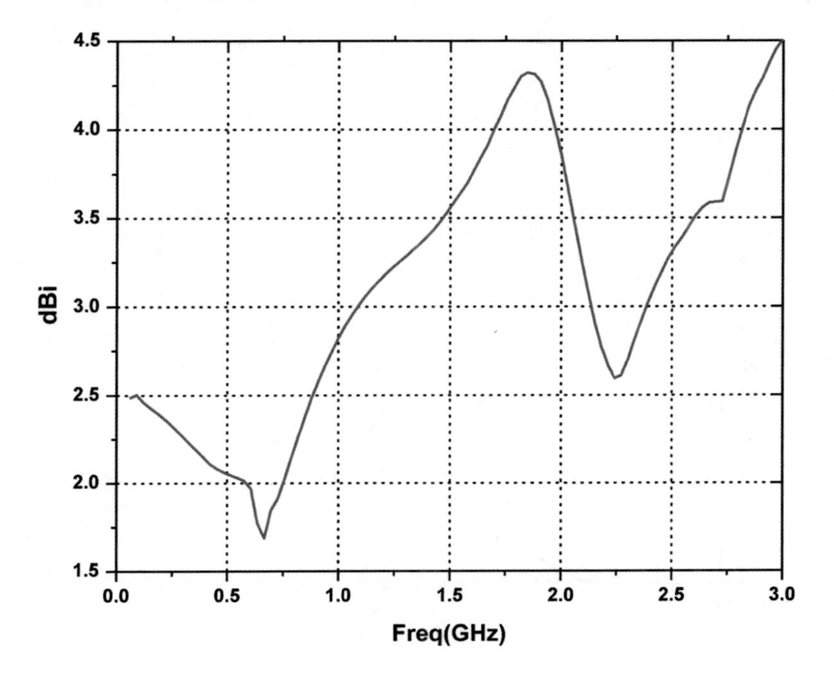

Figure 5. Gain of proposed antenna

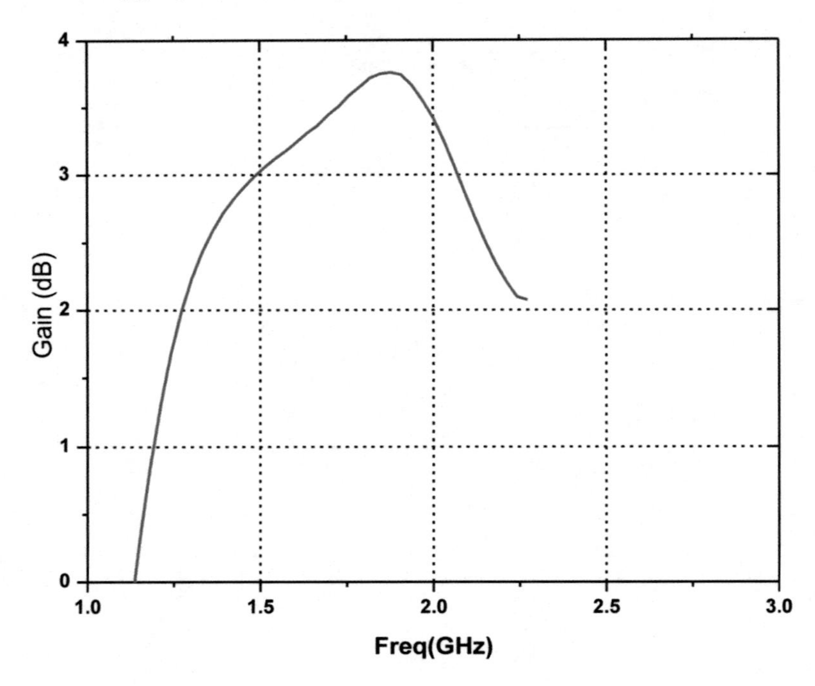

Figure 6. Antenna efficiency of proposed antenna

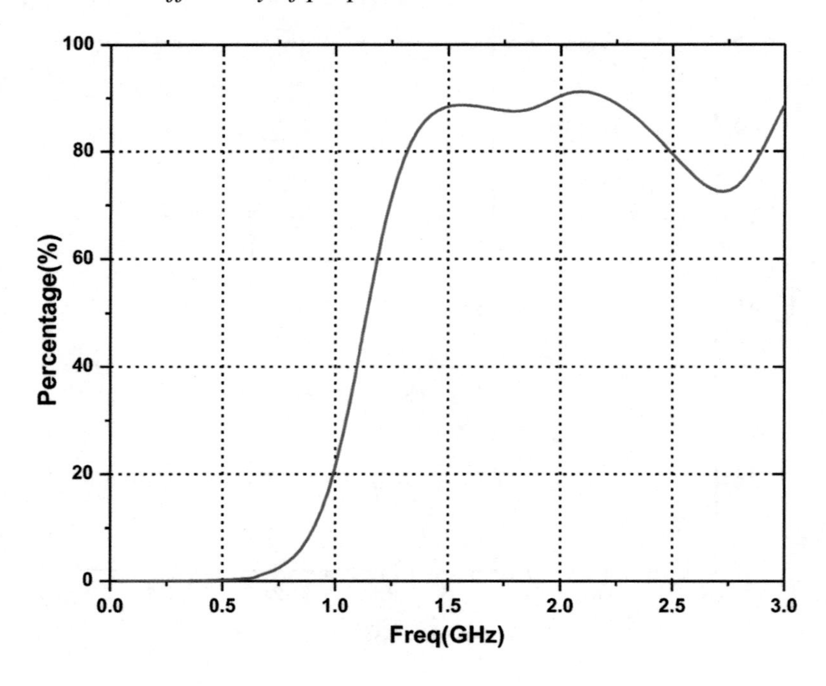

Figure 7. Smith chart

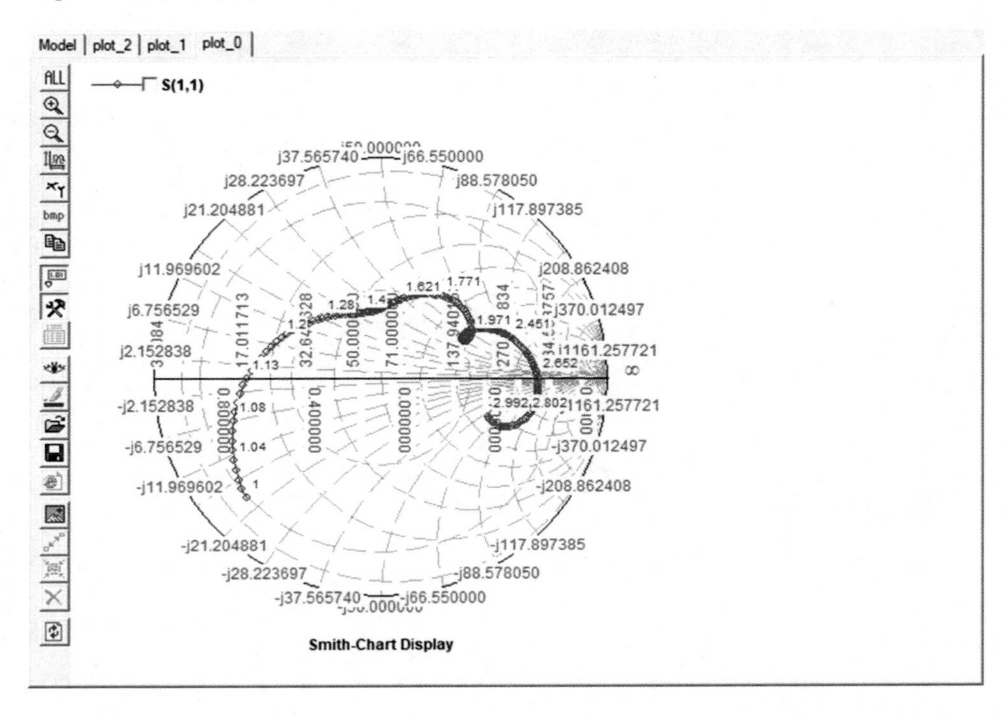

Figure 8. 2D radiation pattern

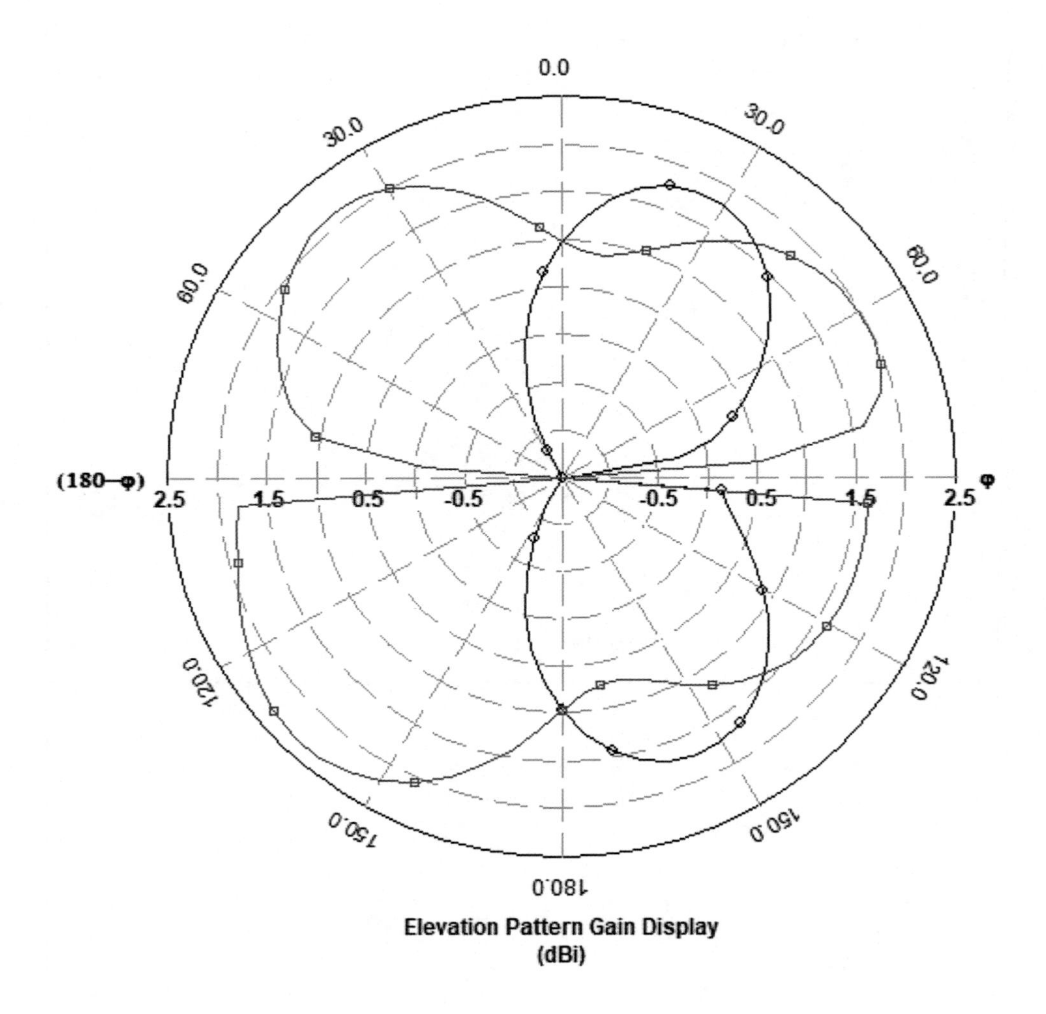

CONCLUSION

A slot loaded 50Ω microstrip fed line rectangular shape microstrip patch antenna has been simulated by IE3D software. The anticipated antenna is designed on glass epoxy substrate which produces a wide bandwidth about 48% and maximum antenna efficiency of about 91%. This antenna is designed to cover the range of frequency between 1.447 GHz to 2.382 GHz which is appropriate for wireless application.

REFERENCES

Chadha & Kumar. (2012). Rectangular Microstrip Patch Antenna Design for WLAN Application using Probe Feed. *IJETAE, 2*(12).

Constantine, A. (1997). *Antenna Theory-Analysis and Design* (2nd ed.). John Wiley and Sons, Inc.

James, J. R., & Hall, P. S. (1989). Handbook of microstrip antennas. Peter Pereginus Ltd.

Mailloux, Mcllvenna, & Kemweis. (1981). Microstrip Array Technology. *IEEE Transactions on Antennas and Propagation*, *29*(1).

Pues & Van de Capelle. (1984). Accurate transmission line model for the rectangular microstrip antenna. *IEEE Microwave* Antenna and Propagation Proceedings, 131.

Singh, Kabeer, Shukla, Singh, & Ali. (2010). Performance Analysis of First Iteration Koch Curve Fractal Log Periodic Antenna of Varying Angles. *Open Engineering, 3*(1), 51-57.

Singh, V. K., Dhupkariya, S., & Bangari, N. (2017). Wearable Ultra Wide Dual Band Flexible Textile Antenna for WiMax/WLAN Application. *International Journal of Wireless Personal Communications, Springer*, *95*(2), 1075–1086. doi:10.100711277-016-3814-7

Yadav & Singh. (2019). Design of U-shape with DGS circularly polarized wearable antenna on fabric substrate for WLAN and C-Band applications. *Journal of Computational Electronics, 18*, 1-7.

Chapter 10
Slotted Wearable Antenna for WLAN and LTE Applications

Nupr Gupta
Bundelkhand Institute of Engineering and Technology Jhansi, India

Rishabh Kumar Baudh
https://orcid.org/0000-0003-1026-2373
Bundelkhand Institute of Engineering and Technology Jhansi, India

D. C. Dhubkarya
Bundelkhand Institute of Engineering and Technology Jhansi, India

Ravi Kant Prasad
Bundelkhand Institute of Engineering and Technology Jhansi, India

ABSTRACT

Slotted wearable antenna is designed at frequency 2.4 GHz due to its application for wireless application and radiolocation. Proposed antenna is used for radiolocation through which detection of objects is possible using a tracking system of radio waves by analyzing the properties of received radio waves. Proposed design employs denim material as a substrate with copper patch as conducting layer. Denim fabric layer of 1mm thickness with permittivity of 1.7 and loss tangent of 0.025 is used as substrate. Dimensions of proposed antenna are calculated using a transmission line model. Proposed antenna has bandwidth percentage of 46% with center frequency 2.42 GHz, and it has high radiation efficiency 93.69%. It covers the frequency range between 2.18 GHz and 3.49 GHz, which works on WLAN applications (2.4-2.484 GHz) and LTE band (2.17 GHz).

DOI: 10.4018/978-1-5225-9683-7.ch010

I. INTRODUCTION

The growth of wearable antenna is enhancing rapidly which is fabricated with clothes, due to low power consumption and its flexibility (Salonen et. al, 2012). Textile antenna finds its different applications in wireless body area networks (WBAN) at operating frequency 2.4 GHz (Samal, Soh, and Vandenbosch, 2014 & Wang et. al, 2012). Due to compact dimensions and low profile it can be easily hidden in garments therefore plays vital role in military applications . Conventional antenna uses hard substrate which creates discomfort to human body, so textile substrates are extensively used presently to fabricate over garments for WIMAX and WLAN (Shahid et. al, 2012). Conductive textiles are used as a dielectric substrate commercially in textile antennas. Bending effect on impedance and radiation characteristics is also studied (Amaro, Mendas, and Pinho, 2011).

In today's electronics world, awareness about location is salient feature of many mobile applications. Radiolocation is used with radar for allocated frequencies of different band for tracking individual as a security feature in different applications. Antenna characteristics are largely affected by its parameters like thickness and permittivity (Roy, Bhaterchya, and Chaudhary, 2013 and Locher et. al, 2006). Textile substrates are characterized by their permittivity and thickness, so it is taken according to applications (Sankaralingam and Gupta, 2010).

A. Antenna Material and Design

Textile antenna is designed at frequency 2.4 GHz using transmission line model. Denim fabric is employed as a nonconductive dielectric substrate while Copper tape is employed as a conductive finite ground plane and radiating metallic patch. Textile antenna mainly consists of denim substrate sandwiched between the conductive layers of copper tape and electromagnetic excitation is given to it using microstrip line feed. Denim substrate is used because of its flexibility, inelastic and planar structure which can be easily worn over garments. Parameters of denim fabric are calculated using a technique explained in (Shahid et. al, 2012) and these parameters are extracted as permittivity $\varepsilon = 1.7$ and loss tangent $\delta = 0.01$.

Rectangular patch has dimensions (47.3×53.8) with three slots cut over it to enhance the bandwidth efficiency of antenna. First slot is cut at the

Figure 1. Geometry proposed of antenna

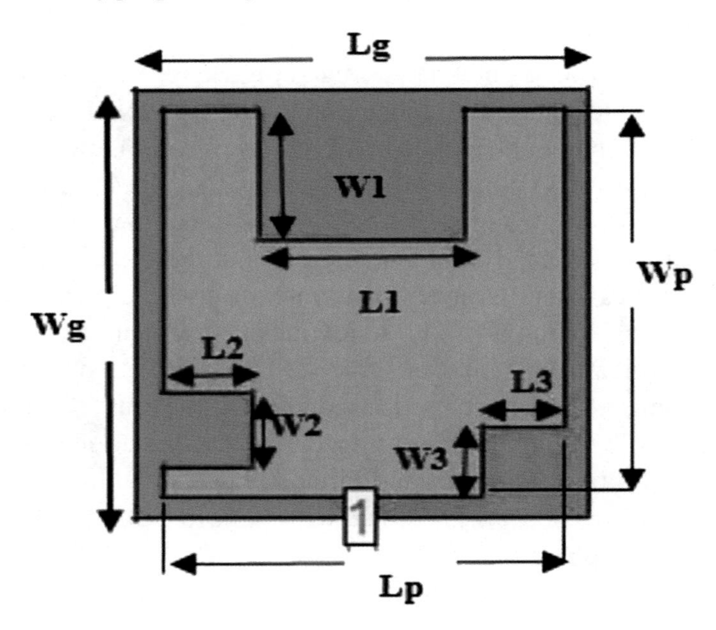

edge of patch and remaining two slots are cut around corners. Microstrip feed technique is used. Feed strip length (L_f) is 3mm, feed strip width (W_f) is 3mm and strip height is 0.5mm. Physical dimensions of proposed antenna which is shown in figure 1are specified in table1.

Table 1. Dimensions of proposed antenna

S.No.	Parameters	Value
1	L_g	53.3 mm
2	W_g	59.8 mm
3	L_p	47.3 mm
4	W_p	53.8 mm
5	L1	24.5 mm
6	L2	9.6 mm
7	L3	10.6 mm
8	W1	18 mm
9	W2	9.6 mm
10	**W3**	**10.6 mm**

The most important parameters which are to be kept in mind while designing an antenna are its length and width which are to be calculated very precisely for designing. The results of patch antenna may be affected by the dimensions of the patch.

In order to design a rectangular microstrip patch antenna

the width and length are calculated as below (Wang et. al, 2012). The width of the rectangular patch may be calculated as

$$W = \frac{C}{2f_r}\sqrt{\frac{2}{\varepsilon_r + 1}} \tag{1}$$

Where, Speed of light (c) is 3×10^8m/s, design frequency (f_r) is 2.4 GHz, dielectric constant (ε_r) is 4.4.

Formula of effective dielectric constant is shown as:

(2)

Where: h= 1.6mm (substrate thickness)

The extension length ΔL is given as:

$$\frac{\Delta L}{h} = 0.412\frac{\left(\varepsilon_{ref} + 0.3\right)\left(\frac{w}{h} + 0.264\right)}{\left(\varepsilon_{ref} - 0.258\right)\left(\frac{w}{h} + 0.8\right)} \tag{3}$$

The length of Patch antenna is given as:

$$L = \frac{C}{2f_r\sqrt{\varepsilon_{ref}}} - 2\Delta L \tag{4}$$

The length (L_g) and the width (W_g) of the ground plane can be calculated by these formulas

$$L_g = L + 6h \tag{5}$$

$$W_g = W + 6h \tag{6}$$

Figure 2a. Conventional wearable antenna

Figure 2b. Return loss of conventional textile antenna

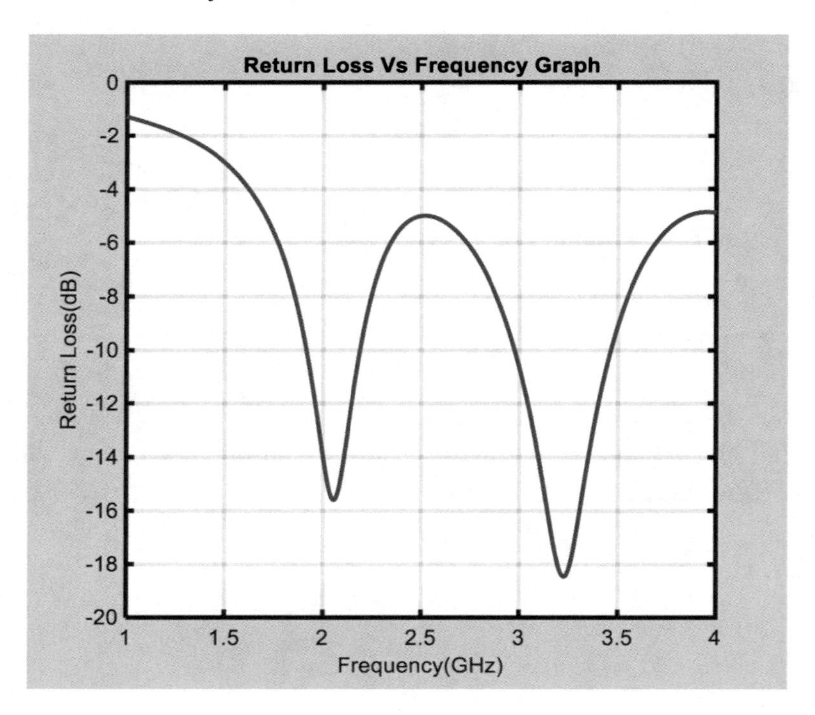

II. FABRICATION PROCESS AND RESULTS OF ANTENNA

GEOMETRY I: - Conventional patch antenna excited with microstrip line

Figure 3a. Single slot antenna

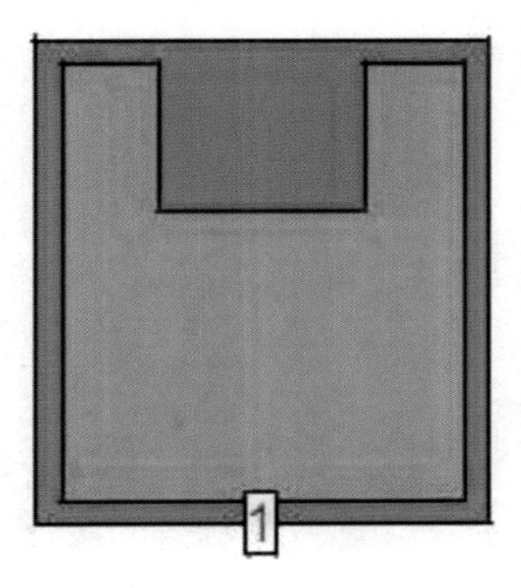

Figure 3b. Return loss graph

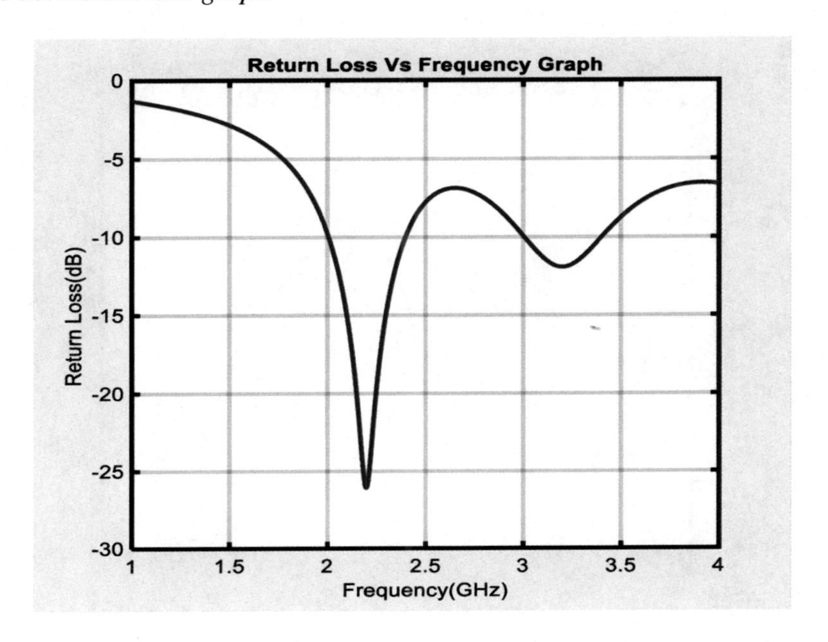

feed gives a dual frequency band. In figure 2(a) design of conventional textile antenna is shown with its return loss displayed in figure 2(b).

Figure 4a. Dual slot antenna

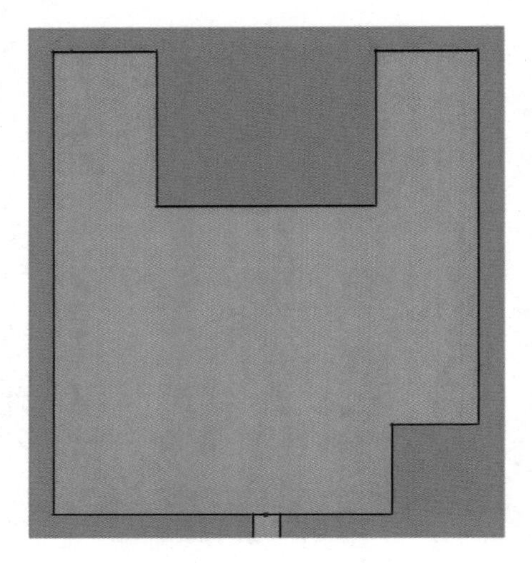

Figure 4b. Return loss of conventional textile antenna

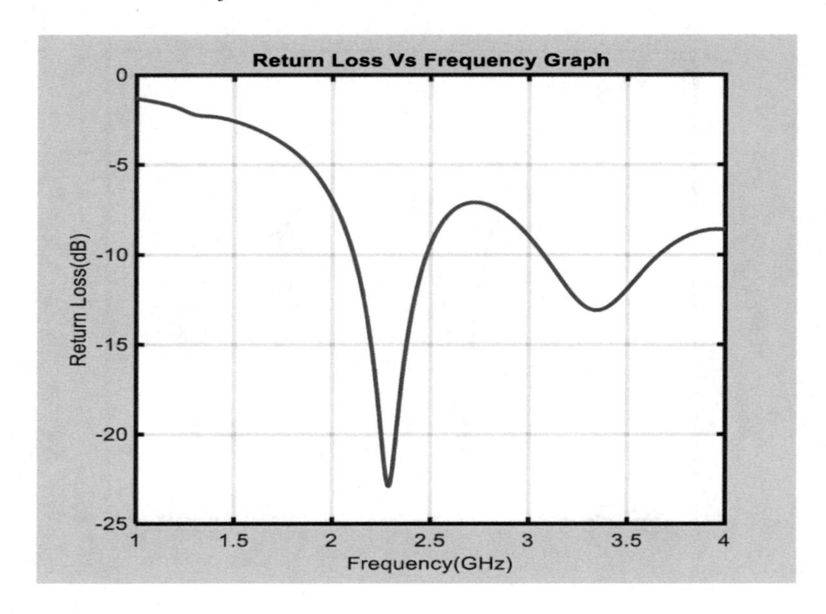

GEOMETRY II: - In this geometry, first rectangular slot is cut at the top edges of patch which is displayed in figure 3(a) which shows improvement in return loss graph as shown in figure 3(b)

GEOMETRY III: - In this geometry, square slot is cut at bottom edge in left side hence it is towards the center which is displayed in figure 4(a). By notching at the edges, the bandwidth of second band is increased which is shown in figure 4(b).

Figure 5a. Proposed textile antenna

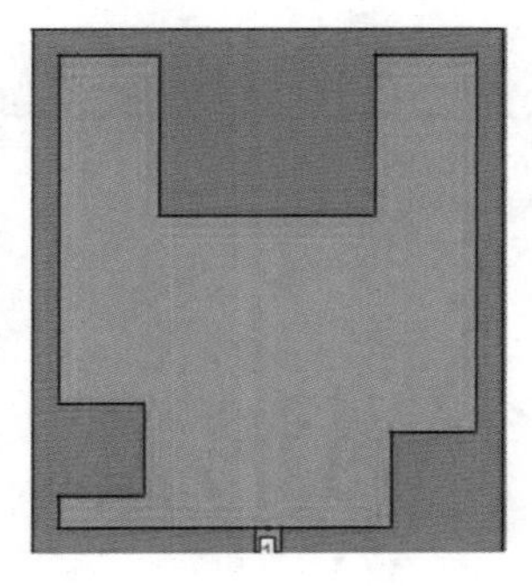

Figure 5b. Return loss graph of proposed slotted antenna

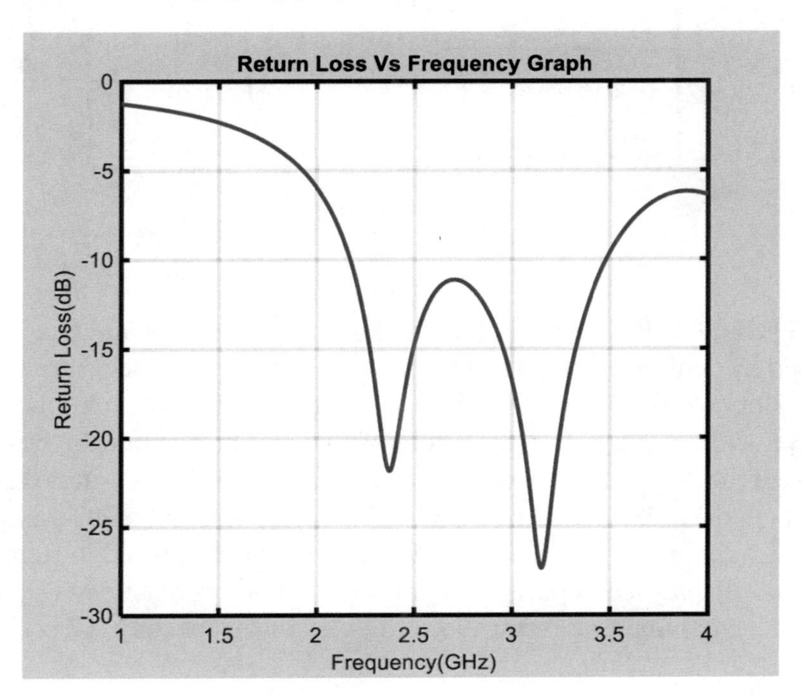

Table 2. Comparison of simulation results of each geometry

Design	Bandwidth %	Maximum Return Loss	Nature of Frequency Band
Geometry I	13.7%	-18.4 dB	Dual band
Geometry II	17.76%	-26 dB	Dual band
Geometry III	16.19%	-22.8 dB	Dual band
Geometry IV	46%	-27.3 dB	Single band

Figure 6. Comparison of S11 parameter of different geometry

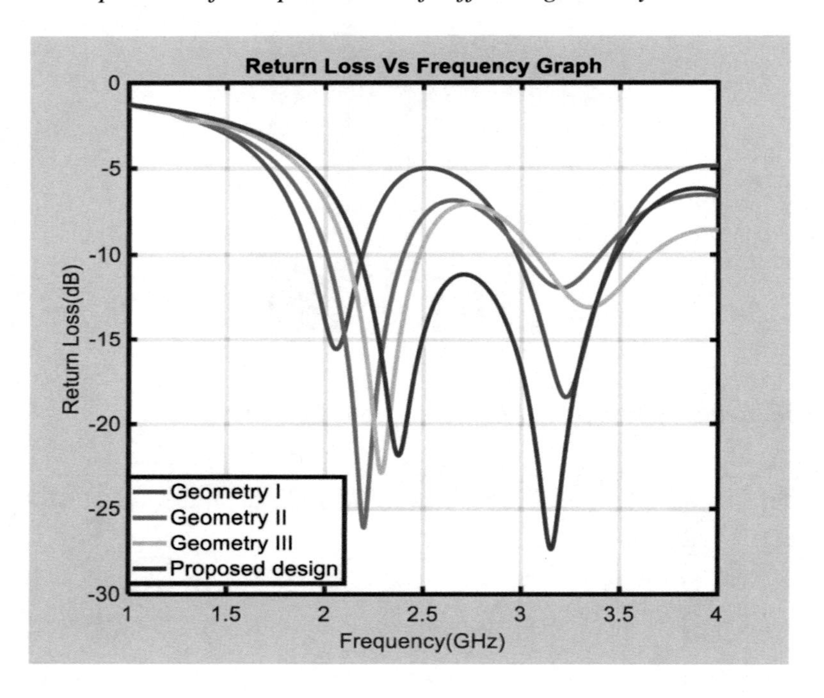

GEOMETRY IV: - This is a final design of proposed antenna which is obtained by cutting a rectangular slot at top edges toward the centre with two square slot cuts of different size at the bottom in toward each other which has been shown in figure 5(a). Effect of this slot brings out the best result because it increases the bandwidth. The Frequency range having return loss beyond -10 dB is 2.18 GHz to 3.49 GHz having bandwidth of 46% and return loss is also improved to -28 dB at 3.2 GHz which is displayed in figure 5(b).

Table 2 display a comparison of various parameters of antenna figure 6 in terms of S_{11} parameter or return loss (dB) against frequency (GHz).

The proposed slotted wearable antenna is large bandwidth antenna analyzed through Zeeland IE3D simulator. The proposed antenna covers frequency range between 2.16 GHz and 3.49 GHz with return loss of -10dB. Simulation result shows maximum return loss of -28 dB at 3.2 GHz. According to analysis of return loss graph, upper cut off frequency is 3.49 GHz and lower cut off

Figure 7. Gain versus frequency graph

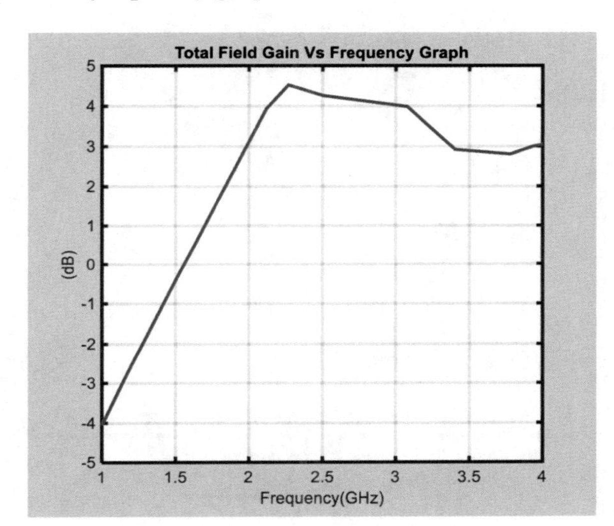

Figure 8. Directivity versus frequency graph

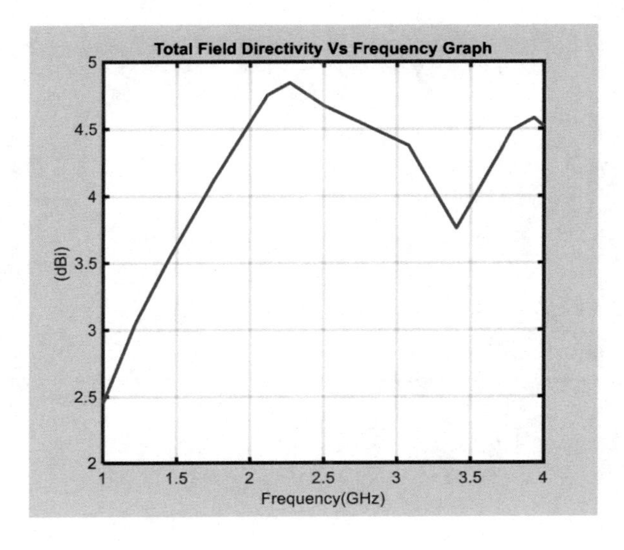

Figure 9. 2D radiation pattern

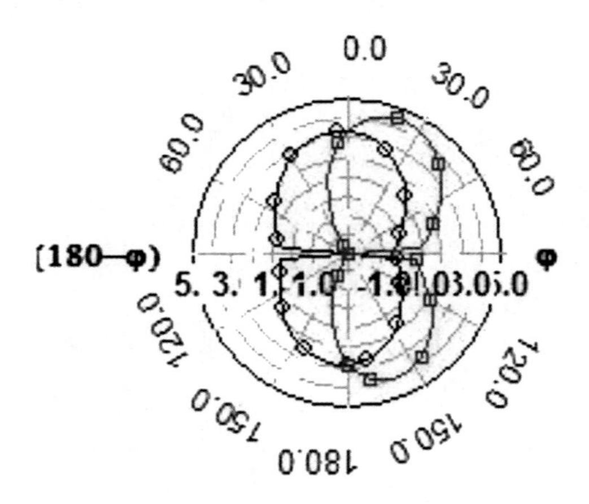

Figure 10. Antenna efficiency graph

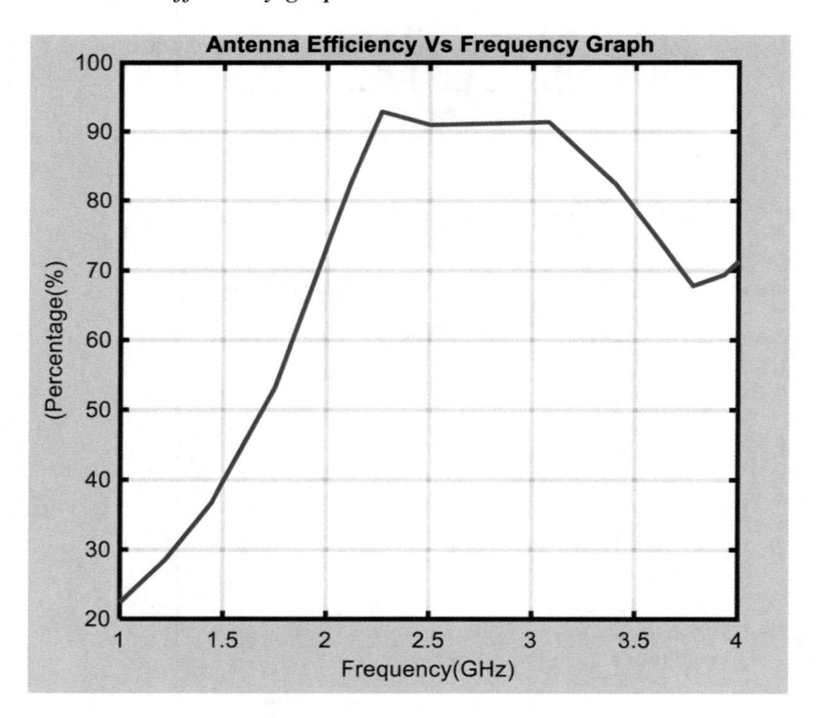

frequency is 2.16 GHz. Bandwidth percentage is calculated as 46% with center frequency 2.42 GHz.

Proposed wearable antenna has gain of 4.36 dBi and directivity of 4.68 dBi at 2.42 GHz which is analyzed by IE3D electromagnetic simulator and results are displayed in figure 7 and figure 8 i.e. gain versus frequency plot and directivity versus frequency plot respectively. Large bandwidth is obtained with respect to conventional (unslotted) textile rectangular patch antenna by slotting and selecting proper line feed technique.

Normalized Pattern of Radiation is simulated to obtain other parameters like gain and half power beam width. Stimulated radiation pattern is displayed in figure 9. 2D radiation pattern is measured at phi 0° and 90°.Propose slotted textile antenna has high radiation efficiency and antenna efficiency at designed frequency of 2.42 GHz. Simulated radiation efficiency is 93.69% and antenna efficiency is 93.05% at 2.42 GHz. Antenna efficiency versus frequency plot is displayed in figure 10.

III. CONCLUSION

Proposed antenna has a high radiation efficiency and antenna efficiency of 93.69% and 93.05% respectively. Slotted wearable antenna has a bandwidth efficiency of 46% centered at 2.424 GHz covering frequency range 2.16 GHz and 3.49 GHz finds its many applications in WLAN and LTE.

REFERENCES

Amaro, N., Mendas, C., & Pinho, P. (2011). Bending effects on a textile microstrip antenna. *IEEE International Symposium on Antennas and Propagation (APSURSI)*, 282-285. 10.1109/APS.2011.5996697

Devi, M., Singh, V. K., Sharma, S., & Bhoi, A. K. (2018). *Antenna for Wireless Area Network and Bluetooth Application, LNEE* (Vol. 462). Singapore: Springer.

Jalil, M. E., Rahim, M. K. A., Abdullah, M. A., & Ayop, O. (2012). Compact CPW-fed ultra-wideband (UWB) antenna using denim textile material. *International Symposium on Antenna and Propagation*, 30-33.

Kaija, T., Lilja, J., & Salonen, P. (2010). Exposing textile antennas for harsh environment. *Milcom 2010 Military Communications Conference*, 737-742. 10.1109/MILCOM.2010.5680300

Klemm, M., & Troester, G. (2006). Textile UWB antennas for wireless body area networks. *IEEE Transactions on Antennas and Propagation, 54*(11), 3192–3197. doi:10.1109/TAP.2006.883978

Kushwaha, R., Singh, V. K., Singh, N. K., Saxena, A., & Sharma, D. (2018). *A Compact Pentagonal Textile Microstrip Antenna for Wide Band Application, LNEE* (Vol. 443). Singapore: Springer.

Lala, K., Lala, A., & Singh, V. K. (2018). *Wide Band Triangular Patch Textile Antenna with Partial Ground Plane, AISC* (Vol. 632). Singapore: Springer.

Locher, I., Klamm, M., Kirstein, T., & Trster, G. (2006). Design and characterization of purely textile patch antennas. *IEEE Transactions on Advanced Packaging, 29*(4), 777–788. doi:10.1109/TADVP.2006.884780

Rahim, H. A., Malek, M. F. A., Adam, I., Juni, K. M., & Saleh, M. I. M. (2012). Basic characteristics of a textile monopole antenna for body-centricwireless communications. *IEEE Symposium on Wireless Technology and Applications (ISWTA)*, 272 - 275. 10.1109/ISWTA.2012.6373859

Prasad, Srivastava, & Saini. (2016). Gain and bandwidth enhancement of rectangular microstrip antenna by loading slot. *2016 International Conference on Innovation and Challenges in Cyber Security (ICICCS-INBUSH)*, 304-307. 10.1109/ICICCS.2016.7542356

Roy, B., Bhaterchya, A. K., & Chaudhary, S. K. (2013). Characterization of textile substrate to design a textile antenna. *International Conference on Microwave and Photonics (ICMAP)*, 1-5. 10.1109/ICMAP.2013.6733490

Sahu, A., Gupta, S., Singh, V. K., Bhoi, A. K., Garg, A., & Sherpa, K. S. (2018). *Design of Permanent Magnet Synchronous Generator for Wind Energy Conversion System, LNEE* (Vol. 435). Singapore: Springer.

Saini, R., Singh, V. K., Singh, N., Saini, J. P., & Bhoi, A. K. (2018). Multi Resonant Textile Antenna with Partial Ground for Multiband Applications. In R. Bera, S. Sarkar, & S. Chakraborty (Eds.), *Advances in Communication, Devices and Networking. Lecture Notes in Electrical Engineering* (Vol. 462, pp. 359–367). Singapore: Springer. doi:10.1007/978-981-10-7901-6_39

Salonen, P., Rahmat-samii, Y., Kamardin, K., & Hall, P. S. (2012). *Wearable antennas: advances in the design characteristics and application* (2nd ed.). Antenna and Propagation for Body-Centric Wireless Communication.

Samal, P. B., Soh, P. J., & Vandenbosch, G. A. E. (2014, January). UWB all-textile antenna with full ground plane for off-body WBAN communications. *IEEE Transactions on Antennas and Propagation*, *62*(1), 102–108. doi:10.1109/TAP.2013.2287526

Sankaralingam, S., & Gupta, B. (2010, December). Determination of dielectric constant of fabric materials and their use as substrates for design and development of antennas for wearable applications. *IEEE Transactions on Instrumentation and Measurement*, *59*(12), 3122–3130. doi:10.1109/TIM.2010.2063090

Shahid, S., Rizwan, M., Abbasi, M. A. B., Zahra, H., Abbas, S. M., & Tarar, M. A. (2012). Textile antenna for body centric WiMAX and WLAN applications. *International Conference on Emerging Technologies*. 10.1109/ICET.2012.6375439

Sharma, P., Yadav, A., & Singh, V. K. (2018). *Design of Circularly Polarized Antenna with Different Iterations for UWB Applications, LNEE* (Vol. 443). Singapore: Springer.

Singh, N., Singh, V. K., Saini, R., Saini, J. P., & Bhoi, A. K. (2018). *Microstrip Textile Antenna with Jeans Substrate with Applications in S-Band, LNEE* (Vol. 462). Singapore: Springer.

Singh, N. K., Sharma, N., Ali, Z., Singh, V. K., & Bhoi, A. K. (2018). *Inset Fed Circular Microstrip Antenna with Defected Ground, LNEE* (Vol. 443). Singapore: Springer.

Singh, R., Singh, V. K., & Khanna, P. (2018). *A Compact CPW-Fed Defected Ground Microstrip Antenna for Ku Band Application, LNEE* (Vol. 443). Singapore: Springer.

Tripati, P., Tomar, U., Singh, V. K., & Bhoi, A. K. (2018). Solution of Economic Load Dispatch Problems through Moth Flame Optimization Algorithm. In R. Bera, S. Sarkar, & S. Chakraborty (Eds.), *Advances in Communication, Devices and Networking. Lecture Notes in Electrical Engineering* (Vol. 462, pp. 287–294). Singapore: Springer. doi:10.1007/978-981-10-7901-6_31

Wang, Z., Zhang, L., & Psychoudakis, D. (2012). Flexible textile antennas for body-worn communication. *IEEE International Workshop on Antenna Technology (iWAT)*, 205-208.

Chapter 11
Missile Structured Wearable Antenna for Power Harvesting Application

Bharat Bhushan Khare
https://orcid.org/0000-0001-8755-9808
UIT RGPV Bhopal, India

Akash Kumar Bhoi
Sikkim Manipal Institute of Technology (SMIT), India & Sikkim Manipal University, India

Sanjeev Sharma
New Horizon college of Engineering, India

Akanksha Lohia
S.R. Group of Institutions Jhansi, India

ABSTRACT

In this chapter, a single element of wearable antenna is designed, and further, to enhance the gain, a wearable rectenna array is designed that can be utilized for the purpose of energy harvesting at 3.14 GHz. The theoretical analysis of received power has been studied. The anticipated antenna array shows the directivity of 8.048 dBi that was used to calculate received power by antenna array at the distance of 10 meters from transmitter. This rectenna array can be used to operate the micro-electronic gadgets and to operate small sensors.

DOI: 10.4018/978-1-5225-9683-7.ch011

INTRODUCTION

Many Industries create some impact on wearable devices. At present time it is the need to keep our self hands free and wireless so the wearable antenna as rectenna fulfills such type of requirement. Wearable rectenna can be the source of a small amount of power to operate some devices like sensors. It is easily wearable and flexible in nature that can be used to harvest the energy. Electromagnetic wave which have high frequency in Gigahertz associated with some amount of energy at the particular frequency so this energy can easily captured by the anticipated antenna and then it is used to operate the device which required the small amount of power (Electronic, 2008; ECC, 2007; Chaudhary, Kim, Jeong and Yoon, 2012).

Here specific transmission frequency is used. There is some standard level of frequency exist in our surrounding like mobile communication frequency range, Bluetooth frequency, radiolocation frequency and at this particular frequency, some amount of power density is available in the environment so we can design the wearable antenna that can operated at particular frequency which able to receive the power and used to operate some microelectronic equipment and gadgets. When the single antenna is to be used, it have some gain in dBm (Decibel milli) but to increase its gain, the combination of multiple antenna is used this arrangement called rectenna array (Naresh and Singh, 2017a; Naresh and Singh, 2017b; Hameed and Kambiz, 2017; Wang, Li, Xu, Bai, Liu, and Shi, 2013; Naresh, Singh, Bhargavi, Garg and Bhoi, 2018a). So here antenna array is designed which have larger gain as compared to single antenna which is operated at four resonant frequency. In this chapter the theoretical analysis is occurred only for 3.14 GHz frequency. According to Federal communication commission (FCC) the range of 3.1 GHz to 3.4 GHz is to be used for radiolocation purpose.

Background

Rectenna is the part of wireless power transmission. The demonstration of wireless power transmission was performed by NASA (National aeronautics and space administration) in 1975. In this demonstration 34000 watts of power was safely transmitted over the 1.5 Km range. In that transmission system, 26 meter long antenna with the 0.5 Megawatt transmitters was used. At that demonstration the large amount of energy was transmitted through antenna this cause the huge radiation so such type of system cannot be established in general use. In this chapter wearable antenna is discussed for energy harvesting (to harvest the received energy of antenna, rectenna circuit is to be used) so this wearable antenna is fabricated on textile material (jeans) and can be fabricated on cloth so this is considered as wearable textile antenna. In this chapter the theoretical analysis is performed to analyze the minimum received energy by wearable antenna, when it is placed at 10 meter distance from radar transmitter.

MAIN FOCUS OF THE CHAPTER

According to Electronic Communication Committee (ECC) report (within the European conference of postal and telecommunication administration) Kristiansand, June 2018 (Electronic, 2008), the range of between 3100-3400 MHz, which is used in radiolocation system, the characteristics of that frequency range has been described in Table 1 in case of radar transmitter. In this chapter the main focus to analyze the minimum power received by wearable antenna at minimum power density (for minimum radar emitted power 55 dBm) of radar transmitter when antenna placed at 10 meter distance from transmitter.

Table 1. ECC report for 3100.3400 MHz

S. No.	Characteristic Name	Min	Typ	Max
1	Radar transmitter emission power (peak power) in dBm (P_e)	55	80	95
2	Antenna gain in (dBi) in antenna main beam (G_e)	25	35	45
3	(P_e+G_e) Peak EIRM according to distance coverage in (dBm)	80	115	140

ANTENNA DESIGN

This antenna is designed on CST microwave studio. This designed antenna has the missile's structure. The microstrip antenna consist of three part named as ground, substrate and patch. The ground of the antenna can be made by non metallic material like mica, textile, PCB (Printed circuit board) etc. But for proposed design, textile material is used for substrate and copper tape is used for ground and patch that makes it flexible. The probe is connected at the lower side of the antenna (Naresh, Singh, Bhargavi, Garg and Bhoi,

Figure 2. Geometry of ground plane

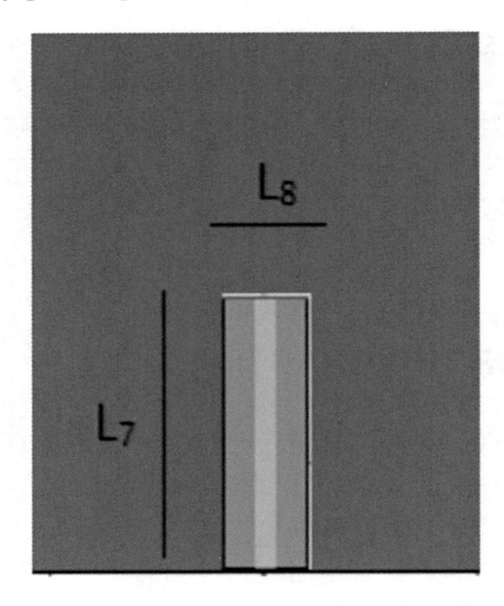

Figure 1. Antenna deign and its geometry

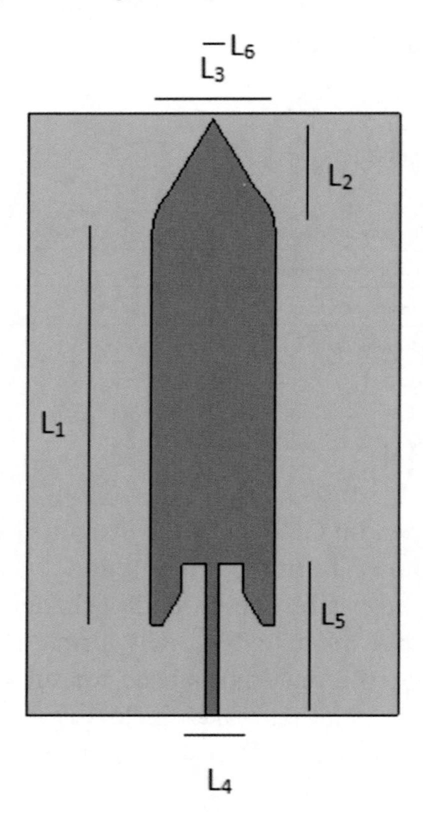

Table 2. Parameter of deigned antenna

Sr. No.	Antenna Parameter	Value
1	Relative Permittivity (ε_r)	1.7 F/m
2	Substrate thickness (h)	1mm
3	Ground plane dimension (L7*L8)	25*7.8 mm*mm
4	Middle length (L1)	65mm
5	Upper length (L2)	19mm
6	Width of antenna (L3)	20mm
7	Inner cut length (L4)	10mm
8	Strip height (L5)	25mm
9	Strip width (L6)	02mm

2018b; Naresh, Singh, Bhargavi, Garg and Bhoi, 2018c). To design this antenna copper tape fabricated on the textile material like jeans cloth. The construction of designed antenna is shown in Figure 1 and Figure 2. The parameter of that antenna is explained in Table 2.

ANTENNA ARRAY DESIGN

To increase the gain of the proposed antenna, the number of same times of antenna is used at same probe then it is considered as antenna array. The gain of this missile's structured antenna is low at 3.14 GHz so the six proposed antenna are connected through the strip (size 152*5 mm*mm) at single probe. The gain of antenna array increased to 8.048 dBi. Antenna array is shown in Figure 3.

ANALYSIS OF MINIMUM POWER DENSITY OF RADAR TRANSMITTER

To analyze the minimum power density of radar transmitter, consider the minimum power density of transmitter and minimum antenna gain according to the ECC report. So from Table 1, the minimum radar transmitter emission power P_e (peak power) in dBm is 55 and antenna gain G_e in dBm is 25. By

Figure 3. Structure of antenna array

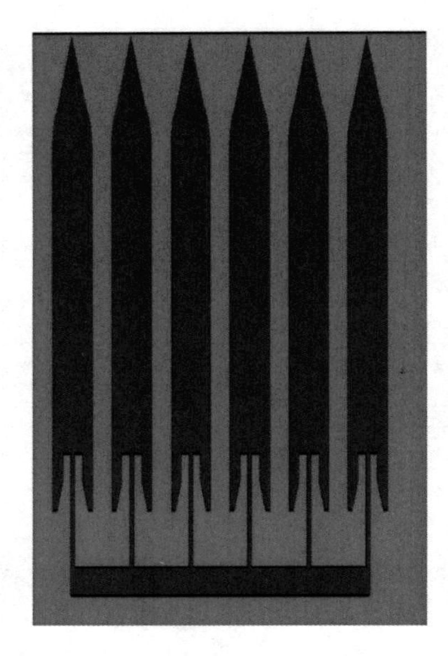

using equation (1), it can be calculated the transmitted power of radar (P_t) and gain of radar (G_e)

$$P_t\left(dBm\right) = 10\log_{10}\left[{P_t}/{10^{-3}}\right] \quad (1)$$

$$G_e = 10^{\left[Ge\left(dBi\right)/10\right]} \quad (2)$$

Power density is the factor of per square meter so when the distance from the transmitter is increase then the power density decreases by the square of the length. Here the power density at the distance of 10 meter is to be calculated.

$$P_d = \frac{P_e * G_e}{4\pi R^2} \quad (3)$$

Where

R is the distance from transmitter and receiver
P_d is the power density
P_t is the minimum radar transmitter emission power
G_e is the antenna gain

Substituting the parameters P_e =316.2277watts, G_e = 316.227 and R = 10 meter in equation (3) we will get the power density P_d = 79.5774 watts/m^2

ACTUAL POWER RECEIVED BY ANTENNA ARRAY

The power actually received by antenna array at the distance of 10 meter from radar can be calculated by

$$P_d = \frac{P_d * G_m * \lambda^2}{4\pi} \tag{4}$$

Where P_r is the actually received by the antenna, G_m is gain of antenna and wavelength of the antenna λ can be calculated by C/f. so the value of λ at 3.14 GHz is 0.0955414 meter. Gain of antenna array is derived by CST microwave studio is 8.048 dBi, so the value of G_m by using equation 2 is 6.3796962. So at 3 meter distance the power actually received by the proposed antenna array by using the equation (4) is

P_r = 0.368776376 W

RECTIFIER CIRCUIT

The received power from array is alternating in nature so to rectify the supply, rectifier circuit is designed. This circuit construct by using schottky diode due to minimum threshold value approx 150 mV. It is designed on Pspice software by using capacitor and inductor. The 1 kilo ohms load is connected on the load side. To define the voltage of received power of antenna consider the 50 ohms source impedance then the value of source voltage can be calculated by

$$V_s = 2 * \sqrt{2 * R_s * P_r}$$

Vs = 12.14539 V

By using this voltage the rectifier circuit is designed on Pspice software which is shown in Figure 4.

Figure 4. Rectifier circuit of proposed antenna array

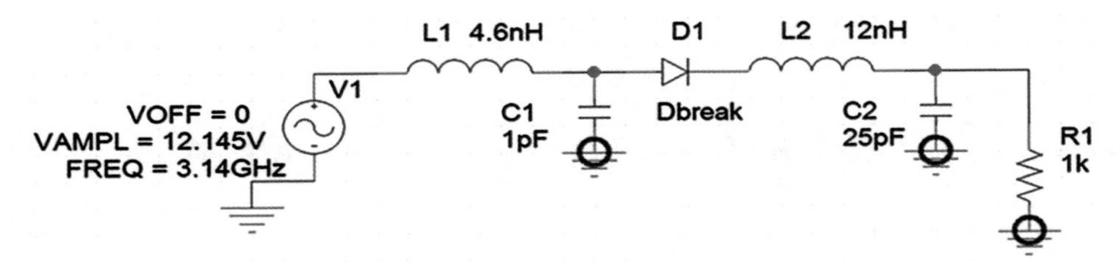

RESULT AND DISCUSSION

Here the theoretical analysis is conducted for receiving the energy which is available at 3.14 GHz in general usage. The range of 3.1 GHz to 3.4 GHz is used for radiolocation purpose and the gain of this antenna at 3.14 GHz I 8.048 dBm and at 3.4 GHz is 6.438 dBm. So the analysis is performed only for 8.048 dBm gain. The antenna array received the power of 0.368776 W which is rectified by the rectifier and used to operate micro-electronic device. This rectenna is fabricated on the textile material so it is easily wearable. The return loss versus frequency graph, predicted by CST software is shown in Figure 5. The output voltage current and power graph with respect to time, predicted on Pspice software is shown in Figure 7 Figure 8 and Figure 9 respectively.

Figure 5. Reflection coefficient versus frequency plot

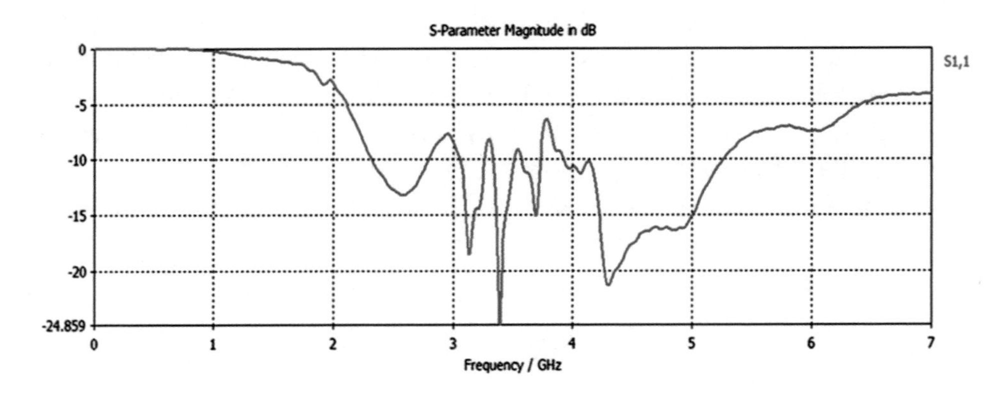

Figure 6. Radiation pattern at 3.14 GHz

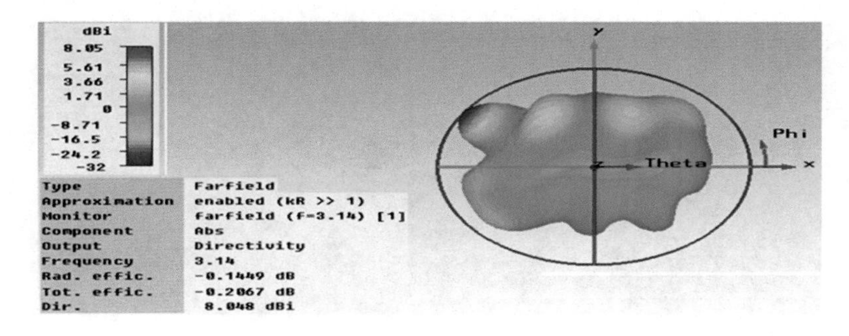

Figure 7. Output voltage graph of rectenna with respect to time

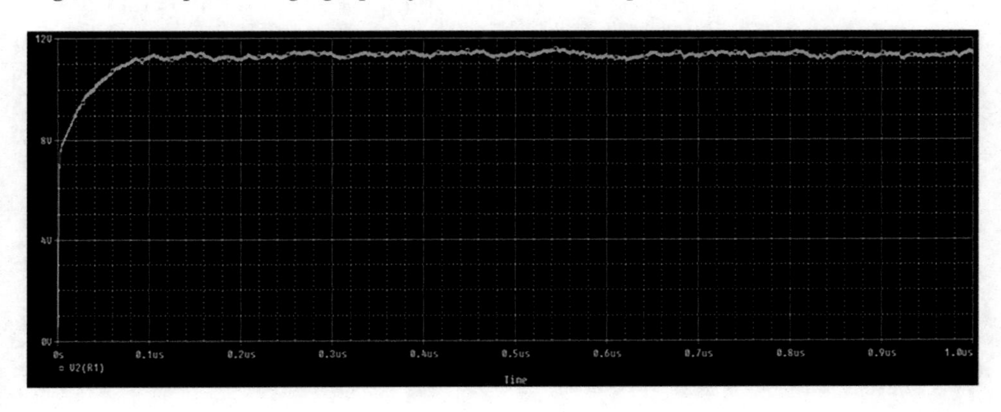

Figure 8. Output current graph of rectenna with respect to time

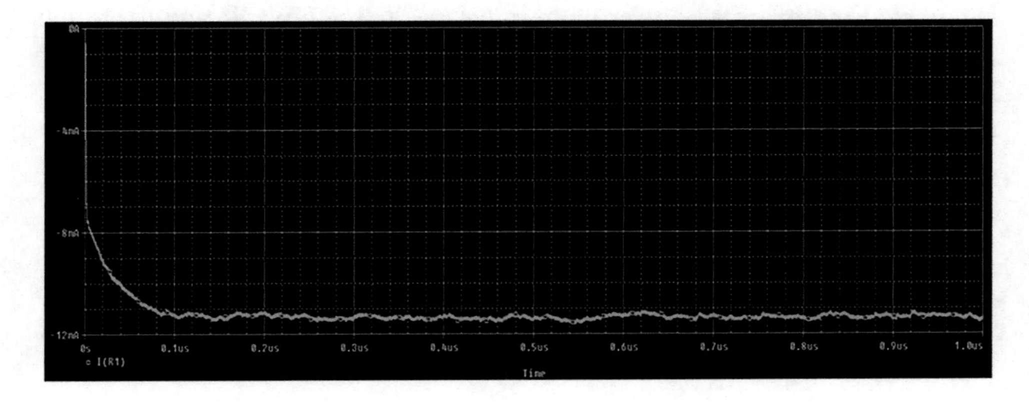

FUTURE RESEARCH DIRECTIONS

Figure 9. Output power graph of rectenna with respect to time

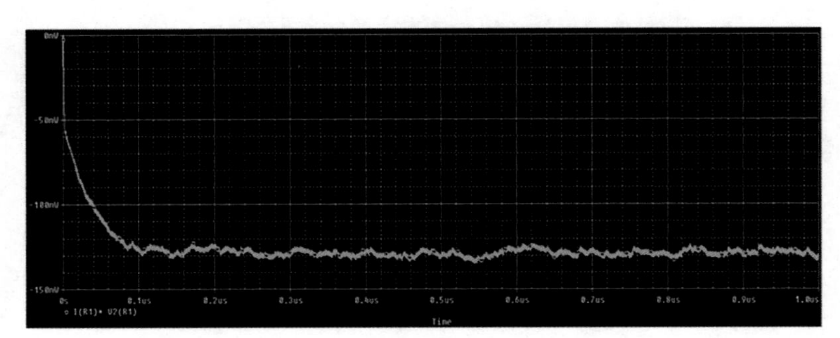

Wearable antenna as rectenna is become popular because it has versatile use like mobile charging without any power supply or battery by using rectenna array and another use to operate the small sensors which required the power in mW. If the gain of the antenna is to be increased by more research then the receiving power is also increased then it required only two or three antenna in array to fulfill the small power requirement without power supply or battery easily. Then rectenna circuits are used more as compared to power bank to charge the cell phones.

CONCLUSION

In this chapter, wearable rectenna array is designed which can be utilized for the purpose of energy harvesting at 3.14 GHz. The theoretical analysis shows that the received power from radar was 0.3687763 W and the value of source voltage received by textile antenna is 12.14 V. the power received by 1 kilo-ohms resistance is 138 mW. This rectenna array can be used to operate the micro-electronic gadgets and to operate small sensors which required small amount of power.

REFERENCES

Chaudhary, G., Kim, P., Jeong, Y., & Yoon, J. H. (2012). Design of high efficiency RF-DC conversion circuit using novel termination networks for RF energy harvesting system. *Microwave and Optical Technology Letters, 54*(10), 2330–2335. doi:10.1002/mop.27087

ECC Report 100. (2007). *Compatibility studies in the band 3400 - 3800 MHz between Broadband Wireless Access (BWA) systems and other services*. Author.

Electronic Communications Committee (ECC). (2008). *Technical Requirement for UWB DAA (Detect And Avoid) devices to ensure the protection of radiolocation services in the bands 3.1-3.4 GHz and 8.5-9 GHz and BWA terminals in the band 3.4-4.2 GHz*. Author.

Hameed, Z., & Moez, K. (2017). Design of impedance matching circuits for RF energy harvesting systems. *Microelectronics Journal, 62*, 49–56. doi:10.1016/j.mejo.2017.02.004

Naresh & Singh. (2017a). Dual band RF Energy Harvester for Wearable Electronic Technology. In *3rd International Conference on Advances in Electrical, Electronics, Information, Communication and Bio-Informatics (AEEICB-17)*. Prathyusha Engineering College. 10.1109/AEEICB.2017.7972428

Naresh & Singh. (2017b). 4.65 GHz Wearable Rectenna for low power Wireless applications. In *International Conference on Electrical, Computer and Communication Technologies (ICECCT-2017)*. Department of Electrical and Electronics Engineering, SVS College of Engineering. 10.1109/ICECCT.2017.8117937

Naresh, B., & Singh, V. K. (2018). Flexible Hybrid Energy Harvesting system to power wearable electronics. In *4th International Conference on Advances in Electrical, Electronics, Information, Communication and Bio-Informatics (AEEICB-18)*. Prathyusha Engineering College.

Naresh, B., Singh, V. K., & Bhargavi, V. (2018). Low Power Circularly Polarized Wearable Rectenna for RF Energy Harvesting. In A. Garg, A. Bhoi, P. Sanjeevikumar, & K. Kamani (Eds.), *Advances in Power Systems and Energy Management. Lecture Notes in Electrical Engineering* (Vol. 436, pp. 131–138). Singapore: Springer. doi:10.1007/978-981-10-4394-9_13

Naresh, B., Singh, V. K., Bhargavi, V., Garg, A., & Bhoi, A. K. (2018). Dual-Band Wearable Rectenna for Low-Power RF Energy Harvesting. In A. Garg, A. Bhoi, P. Sanjeevikumar, & K. Kamani (Eds.), *Advances in Power Systems and Energy Management. Lecture Notes in Electrical Engineering* (Vol. 436, pp. 13–21). Singapore: Springer. doi:10.1007/978-981-10-4394-9_2

Wang, X., Zhang, L., Xu, Y., Bai, Y. F., Liu, C., & Shi, X.-W. (2013). A tri-band impedance transformer using stubbed coupling line. *Progress in Electromagnetics Research*, *141*, 33–45. doi:10.2528/PIER13042907

Chapter 12
Wearable Antenna Materials

Vimlesh Singh
https://orcid.org/0000-0002-5758-0603
MRIIRS, India

Priyanka Bansal
MRIIRS, India

ABSTRACT

The aim of this chapter is to identify various materials being used currently for antenna design classified as wearable materials. In the current scenario, no study was found where collectively all the available materials and their properties could be discussed including their pros and cons features. This chapter identifies various available materials on the basis of their characteristics, availability, and the methodology of fabrication being involved along with their corresponding properties. Post detailed study and analysis done in this research enabled us to broadly classify the materials as conductive and substrate materials. This brings to the understanding that earlier no such broader classification was made available, and hence, a comprehensive study would provide us better information availability on various wearable antenna materials.

INTRODUCTION

With the advancement in communication technology, the application of antennas has evolved in many innovative concepts. One such concept in recent times is a wearable antenna. The application of antennas involving sensors merging with human use products is defined as wearable antenna.

DOI: 10.4018/978-1-5225-9683-7.ch012

These can be in the form of clothing or fabric or tech gadgets like the smart watch, tech specs etc. The mobility and personalized aspect of a wearable antenna make it one of the most profound innovations of modern science in the field of communication. Hence it finds its diverse application in the form of remote monitoring, tracking & navigation, public safety etc.

E textile is one such wearable antenna where the sensors are integrated within the wearable fabric to be used by humans for various purposes and objectives. Usage of e textiles in medical facility or at the times of disaster management etc makes it perhaps one of the most important wearable antennas because of the scale of purpose it aims to solve. Whether it address the lifestyle diseases or helps it in its management or taking the healthcare treatment to an entirely new dimension, the vitality of e textile can neither be ignored nor disputed.

Like any other technology, E textiles do face challenges, in terms of environment changes or life style changes happening in the modern world and even unpredictable issues like the birth of diseases never heard before, etc. Therefore, it is imperative to not only understand the E textile in its essence form but also how various materials would make a difference in combating and addressing issues for which wearable antenna was conceived for.

WEARABLE SYSTEM

Wearable system can be designed as per requirement of application. For wearable system process flow is shown in figure.1. The data generated by a wearable device can be processed on line as well as offline. The most benefitted generation from these wearable is the young generation, which can justify with their dual responsibilities of child and parents. To interpret the raw data generated by a critical wearable device, we need a sequence of actions that keep on adding value to the raw data.

- Sense
- Analyze
- Store
- Transmit
- Utilize

Let us consider an example of a heart patient. To check the criticality of his own condition he will visit a physician, where a nurse documents the vital data

Figure 1. Flow chart: Data flow between transmitting end to receiving end

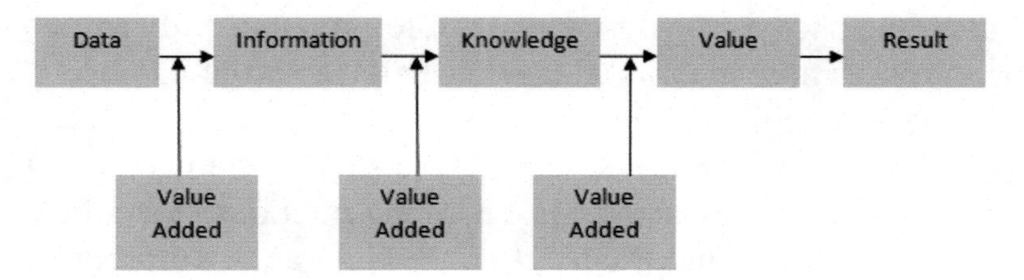

collected using instruments (blood pressure monitoring, ECG, EKG). Thus the raw data from the body is converted in to information in terms of heart rate and blood pressure. Now this information is forwarded to the physician who further add value to the information in terms of knowledge gained by him over the time and experience. Thus this information is converted in to knowledge i.e. the person will get to know what is his medical condition and what is the medicine prescription for him. After taking proper medicine prescribed by the physician, patient can get the results (health improvement).

WEARABLE MATERIAL CLASSIFICATION

Wearable technology is smart clothing that embeds electronic component in fabric resulting in a "conductive yarn". This is integrated in the textile by knitting, weaving, embroidering, printing, sewing, braiding, nonwoven textile and chemical treatment forming the basis of what is termed as e-textile. The yarn of the textile hence senses and responds as per the surrounding conditions. These conditions of the environment are predominantly governed by factors such as pressure, temperature, thermal behavior etc. Based on the nature of the responses the E textiles can be divided into 3 categories, namely

- Passive Textile
- Active Textile
- Proactive Textile

Passive Textile

In this kind, the textile senses the surrounding condition with the help of sensor(s) integrated in the woven fabric. Since the textile only undertakes the received information of the surrounding and no other activity is attached with the mechanism, it is termed as Passive textile. However the received signal could be used for analysis and its interpretation and can be used for various purposes and benefits. Example, Passive textile can be effectively used in medical facility, as an e textile for hospital clothing like patient's gown(s). The data thus received during the stay of the patient could bear numerous characteristics and physical representations. Like ECG signal (Coosemans, Hermans and Puers, 2006) reading and interpretation, EMG signal (Linz, Gourmelon and Langereis, 2007) and EEG signal (Lofhede, Seoane, Thordsttein, 2010) or movement of muscle movement (Meyer, Lukowwicz and Troster, 2006) etc.

Active Textile

These sensors integrated in an active textile works both for sensing and reacting. In addition to mere sensing and receiving the signal like the passive textile this also reacts to the surroundings in the form of transmitting signals and hence providing information to the end server. This is done with the composite unison of a radio frequency element, harvesting device, power generation and human body interface system (Baurley, 2004). This can be used as a semi automatic system and deployed for various purposes like tracking and navigation, healthcare, rescue mission especially during natural calamity etc

Proactive Textiles

The most advance form of E textiles come under Proactive Textile as the sensor(s) in them sense, react and adapt with the surrounding conditions (Edmison, Jones, Nakad and Martin, 2002; Baurley, 2004) and further still can be programmed for each individual separately. This is a full automatic system. Because of its automation property it finds its usage in numerous fields.

Wearable Antenna Material parameter: Design & development of wearable antenna is different from conventional antenna design approach because its fabrication needs to use body friendly materials. In this regards it is important to analyze electromagnetic properties of materials. At micron level properties of microwave material depends on its mode of propagation. So for design of wearable antenna, electromagnetic force is developed in following ways:

1. Field force developed due to static charge. It is due to fundamental particles.

2. Field force developed due to kinetic energy of charge particle migrates (skin depth).

Macroscopic level characteristics of microwave material extremely depend on following parameter:

1. Basic parameter: Permeability (μ), Dielectric constant (ε) & Conductivity of material (σ)
2. Medium of antenna (linear or non-liner/ homogenous or heterogeneous/ isotropic or anisotropic)
3. Material behavior toward frequency of application

Macroscopic properties of textile material explain behavior towards signal transmission, reception and physical reaction of it to human body interface. So to provide solution towards wearable technology it is necessary to know materials being used for it.

WEARABLE MATERIALS

Wearable antennas material selection should largely be biocompatible, flexible, stretchable and conducive to human comfort. Therefore geological, tropical and local conditions also add to the kind of material being selected. In normal practice two types of materials are used in wearable technology. The first one is a conductive surface that forms a ground plane and a radiating surface. Second material is dielectric which works as platform / mechanical support for radiating patch. Substrate material also moderates the electrical characteristics of wearable antenna by extracting total actual power of radiation. The characteristics of wearable antenna is its dielectric constant, thickness of laminate, thermal coefficient, dissipation factor, and many other tradeoffs. Any shrink in dimension of a wearable changes the parameter of substrate material like structural and tensile strength, resistance to chemical, impact resistance, flexibility, bonding, machine ability, and many others. A very wide range of substrate material of sapphire, alumina, PTFE (polytetrafluorethylene), polyphenylene, semiconductor substrate rutile polystyrene, & ferromagnetic used for different applications wearable application. The choice of substrate material depend on its application because there is no ideal substrate material.

Figure 2. Flow Chart: Classification of Materials

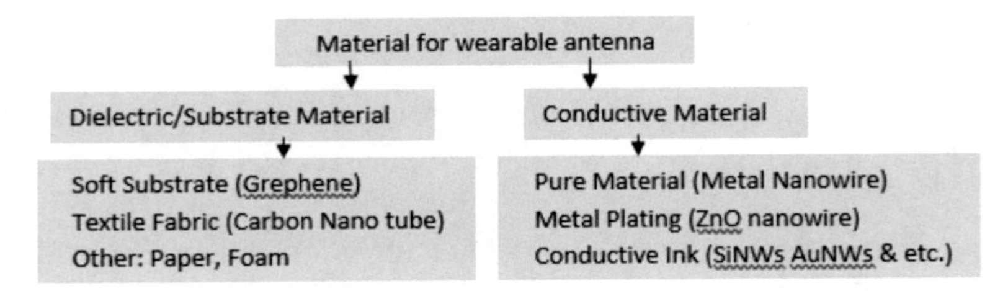

Ceramic Material

Ceramics are preferred as wearable textile material because of thinness and light weight so it has durability as well as usage without any discomfort. Polyimide is one of most widely used substrate material which has high temperature stability as well as chemical resistance. Researcher also reported that polyimide has temperature stability of 3600°C to 4100°C which is in range of glass transition temperature (Rim, Bae, Chen, Marco and Yang, 2016). This high thermal stability support slows down performance of device with time. In same category poly dimethylsiloxane (PDMS) is recently developed ceramic material which has easy processing, remarkable elasticity, transparent appearance, high thermal and chemical stability. Besides artificial materials some natural materials have been developed as substrate materials for wearable technology. Silk fiber is one of more promising solution of wearable technology because it's a biomaterial, attractive and most importantly shows resistance toward mechanical deformation (Li et al., 2016). In future increase in demand of wearable electronics will lead to the development of biocompatible, biodegradable, bio integrated solutions in terms of lightweight, inexpensive, flexibility for RF solution to communicate and energy harvesting (Kim, Kim, Wu, Liu, Kim, Huang, Hwang and Rogers, 2009; Zhu, Wang, Leow, Cai, Loh, Han and Chen, 2016).

Semiconductor Substrate

For wearable textile passive /active circuits design, semi-insulating materials are preferred because they are high resistivity semiconductor materials. Graphene is most promising material for design of wearable and flexible device because it is an active material which support deign of active and passive both type of circuits due to high carrier mobility, good optical characteristics,

Table 1. Substrate material for wearable application (Nathan et al., 2012)

S. No.	Substrate Material	Appearance	Thermal Characteristics	Chemical Characteristics & Moisture Absorption	Cost	Temperature Stability
1	Polycarbonate (PC)	Clear	Poor Coefficient Thermal Expansion (CTE)	Good chemical resistance	Inexpensive	155⁰C
2	Poly ethylene Napthalate (PEN)	Clear	Moderate Coefficient Thermal Expansion (CTE)	Good chemical resistance & Moderate moisture absorption	Inexpensive	150⁰C
3	Polyester (PET	Clear	Moderate Coefficient Thermal Expansion (CTE)	Good chemical resistance & Moderate moisture absorption	Inexpensive	120 ⁰C
4	Polyimide (PI)	Orange color	high Coefficient Thermal Expansion (CTE)	good chemical resistance & high moisture absorption	expensive	275⁰C
5	Polyethersulfone (PES)	Light yellow	Moderate Coefficient Thermal Expansion (CTE)	good dimensional stability, poor solvent resistance, expensive, moderate moisture absorption	expensive	230⁰C
6	Polyetheretherketone (PEEK)	Amber color	Moderate Coefficient Thermal Expansion (CTE)	good chemical resistance & low moisture absorption	expensive	250⁰C

mechanical strength, low cost of production (Allen, Tung, and Kaner, 2010; Novoselov, Ffal, Colombo, Gellert, Schwab, and Kim, 2012). Graphene is a 2-dimensional macrostructure of carbon atom in layered form to achieve acceptable electromechanical characteristics. In (Pereira, Neto, and Peres, 2009; Huang, Pascal, Kim, Goddard, and Greer, 2011) researcher reported elongation of carbon lattice by crystalline graphene flake with negligible defect in it. Author reported that a very slight gap increase strain of 23% in material that make material flexible.

Graphene characteristics are exploited by introducing defect and flakes in material itself. For preparing Graphene wafer chemical vapor deposition technique is used that increases strain in material (Lee, Bae, Jang, Jang, Sim, Song, Hong and Ahn, 2010; Fu, Lioa, Zhou, Zhou, Wu, Zhang, Jing, Xu, Wu, Guo, and Yu, 2011). With change is size of wafer its strain bearing capacity is changed. The fractured Graphene structure show improved piezo sensitivity because of variation in rate of contact. Yao, Ge, Wang, Wang, Hu, Zheng, Ni, and Yu, (2013) reported that fractured microstructure of Graphene show very high pressure sensitivity, these polyurethane structure are nanosheets. This fractured Graphene composition show 95% strain deformation characteristics and fast conduction path for wearable application. Research also explain graphene & copper mesh based fabric developed by chemical vapor deposition work as tactile sensor for human motion monitoring (Wang, Wang, Yang, Li, Zang, Zhu, Wang, Wu, Zhu, 2014), torsion sensor (Sun, Keplinger, Whitesides, and Suo, 2014; Tao, Hwang, Marelli, An, Moreau, Yang, Brenckle, Kim, Kaplan, Rogers, and Omenetto, 2014) and acoustic signal acquisition (Wang, Yang, Lao, Zhang, Zhang, Zhu, Li, Zang, Wang, Yu, 2015).

The ultrahigh sensitivity of macro graphene structure relates with synergistic effect of material so it is used for position-control due to high interfacial resistance between copper mesh and locally oriented zigzag cracks (Yang et al., 2015). Another method to obtain high sensitivity is to develop quasi – continuous nano graphene wafer (Li et al., 2016a). A graphene based stacked

Figure 3. Graphene with Copper mesh fabrics and electromechanical property (Yang, Wang, Zhang, Li, Shi, He, Zheng, Li, Zhu, 2015)

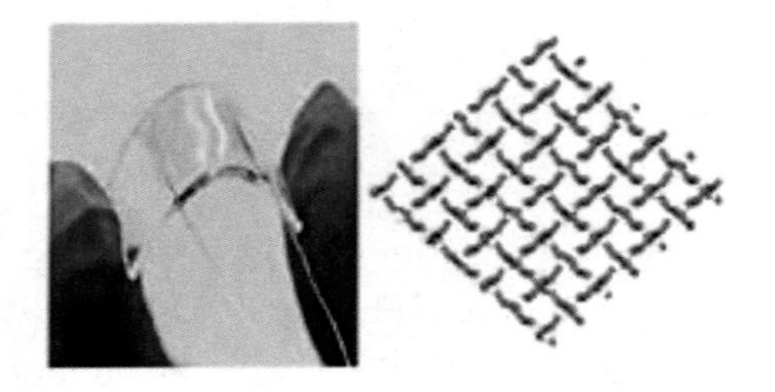

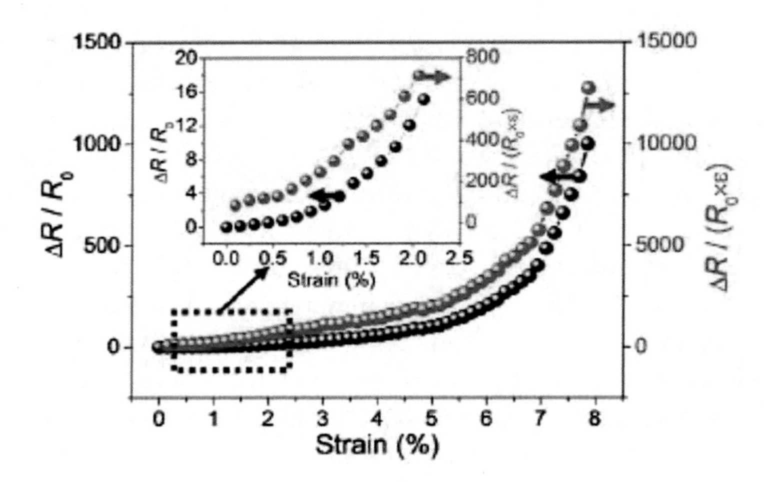

percolative film was reported (Hempel, Nezich, Kong and Hofmann, 2012) for high sensitive application. Textile material was developed by spraying layer of graphene liquid on flexible substrate on plastic in the form of thin wafer of overlapped flakes. This overall effects the sensitivity of material with change in size of wafer, deposition technique and number of graphene material flakes. As graphene achieved optimal flexibility & greater conductivity this can be used as electrode material with sensing element, also as interconnect for specific type of sensor. This technology makes use of the piezoresistive principle where pressure stimulus changes resistance by variation in contact between two same face-to-face electrode (Li et al., 2016b). The different type of smart hybrid textile materials developed by using graphene design of wearable tactile have been reported. A new combination of composite nanofibers of Carbon Nano tube and graphene were reported (Lee et al., 2016), this material maintain constant pressure for bending with a radius up to 80 mm.

Carbon Nano Tube (CNT)

Carbon has unique characteristics that is its allotropic property means its chemical property remain same but physical representation is different. Carbon nano tube is one-dimensional structure of carbon atom in cylindrical nanostructure structure orientation, which can be used as active materials for development of tactile sensors or wearable antenna design. These nano structure show significant charged carrier mobility, optimal chemical stability and robust mechanical characteristics (Hu, Hecht and Gruner, 2010; Yamada et al., 2011). Carbon nano tube show great sensitivity at nanoscale towards mechanical deformation if suitable value of chiral angle is selected.

Carbon nano tube hollow structure absorb extra energy that show exceptional elastic properties with elevated tensile strength of approximately 40% strain (Obitavo, Liu and Sens, 2012). It is very challenging to develop well-controlled growth of Cabon Nano tube with specific chirality, so development of individual Single Wall Nano Tube is not possible at large-scale. For design of carbon nano tube based wearable material lithography and printing technique are used (Hammock et al., 2012).

In carbon nano tube intrinsic and intertube resistance work for strain sensing of carbon nano tube. In both resistance intertube resistance is more crucial for electromechanical performance of CNT network. To make composite, CNT mix with elastomer to form thin film, that can be used for wearable application. Nonlinear piezoresistive characteristics of CNT was observed

Figure 4. Carbon Nano Tube; (a) Schematic of Intrinsic & Inter-tube resistance in CNTs network; (b) CNT-graphene hybrid model with structure strength (Yamada et al., 2011); (c) Stage wise morphology of CNT film with its stretching electrical response (Yamada et al., 2011)

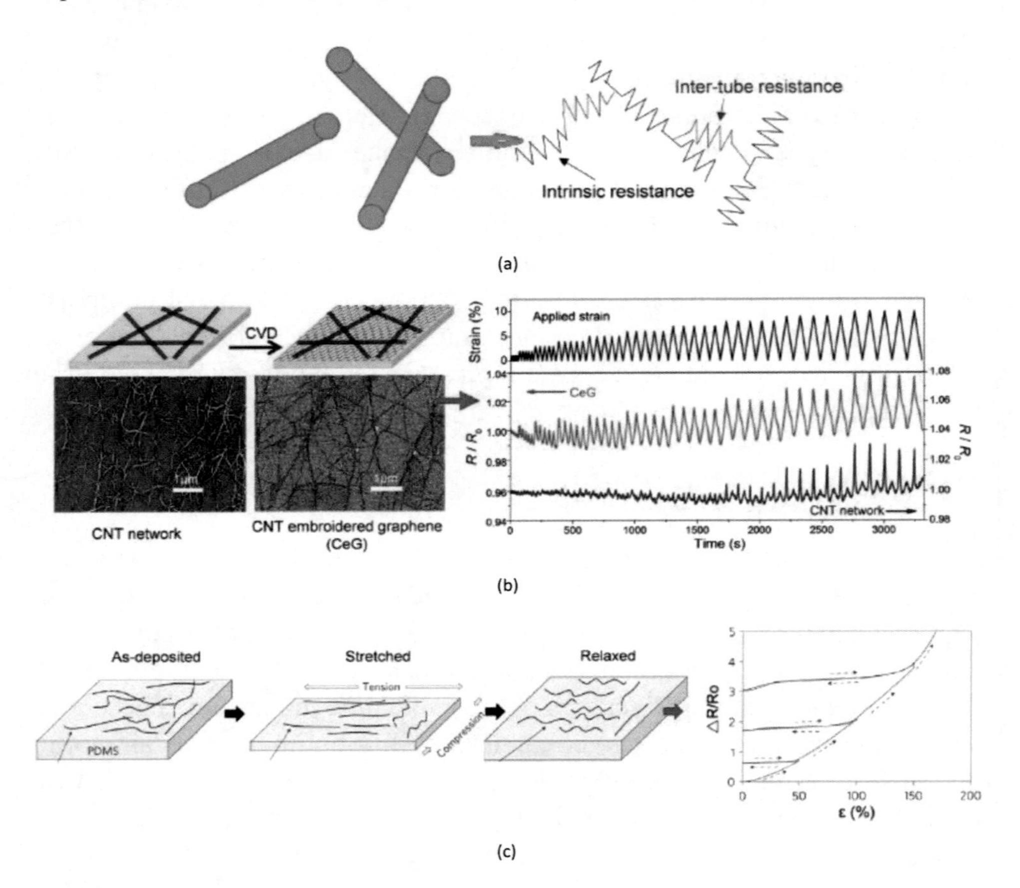

for percolation threshold of approximately 20 in glass film (Hu, Karube, Yan, Masuda, and Fukunaga, 2008). In design of CNT network Interfacial interaction between Carbon nano tube and polymer work as strain sensors. In case of weak delamination in interfacial interaction causes electrical failure in wearable material over the time. It was reported earlier that carbon nano tube deposited directly to the surface of super elastic material work as mechanical sensor (Zhu and Xu, 2012). A coating of carbon nano tube network film on 3 M VHB 4905 substrate show increased resistance at 100% strain that also maintain 700% electrical conductivity (Hu, Yuan, Brochu, Gruner, and Pei, 2009). The loss in conductivity is often observed due to stretch and cracks in CNT network. So to overcome instability in structure, CNT networks are developed on polymer. CNT and grapheme hybridization structure resist to bundling and buckling deformation of wearable material (Chen, Wu, and Yao, 2004). In individual CNT structure and CNT network electromechanical characteristics are different from assembly of CNT on film, yarn or fiber. The CNT network laminated pressure sensitive rubber array work as multifunctional artificial skin.

CONDUCTIVE MATERIAL

Conductive materials have been extensively used for fabrication of wearable textile. Several conductive materials are conductive elastomer composite, ionic conductors, conductive polymer and fluid embedded in elastomeric channels. Conductive material composite are composite of insulated material enriched by conductive fillers. In normal practice resistive-type force-sensitive materials of low fabrication cost are preferred.

The piezoresistivity behavior of conductive material composite primarily occur due to:

1. Break-up produced due to strain and reformation of percolation path.
2. Intrinsic piezoresistivity of material due to change of band structure of material.
3. Variation in resistance of material in electron tunneling.

In conductive material CNT (Hu, Karube, Yan, Masuda, and Fukunaga, 2008), conductive organic (Choong, Shim, Lee, Jeon, Ko, Kang, Bae, and Lee et al., 2014), nanowire (Huang, Xiao, and Fu, 2015), carbon black (Das, and Khastgir, 2002), graphene (Stankovich, Dikin, Dommett, Kohlhaas, Zimney, and Stach et al., 2006), metal particle (Hyun, Park, Park, Kim, Xia, Hur, and Kim et al., 2011) carbon fiber (Chen et al., 2004) are used as filler. The performance parameter of conductive material depends on filler material, concentration, dispersion, geometry, modulus and morphology of the filler materials. Tunneling current between particles of wearable are promoted by filler particles of sharp protrusion (Ma, et al., 2015). Strong bonding force between filler particles and polymer are required to enhance stretchability, hysteresis and cyclic stability performance (Amjadi, Yoon and Park, 2015). For mass production of wafer uniform characteristics and consistent allocation of conductive fillers material in matrix is explored (Jurewicz, et al., 2011). To detect any kind of mechanical deformation porous structure are introduced in material (Jung, et al., 2014).

Active material used for design of wearable material has concentration of filler approximately equal to percolation level, this also help to determine piezoresistive property of material. The major challenges associated with conductive composite material are its sensitivity with temperature, environment and creep. In case of organic conductive polymer material, they are fabricated at low cost for large area by roll –to roll fabrication technique and screen printing. Now a days conductive polymer are preferred because they are optimal filler materials for conductive elastomer composites. So that they can be directly mounted on active electrode element in certain type of pressure sensors, where active material is sandwiched between electrodes. An ultra sensitive pressure sensor was reported for wearable application (Russo et al., 2011; Pan et al., 2004). This sensor was prepared by hollow sphere microstructure of poly pyrrole hydrogel as active element. In these active element copper foil work as top electrode and Indium tin oxide based conductive terephthalate (PET) film work as bottom electrode. To prepare wearable device organic semiconductor polymer is preferred because of fast carrier transport of silicon material. Also it has good flexibility and stretchability characteristics towards the conventional plastic. For high carrier mobility formation of crystalline structure of material must have highest degree. This increase deformability of material and in that case nanoconfinement provide solution.

Nano confinement are semiconductor polymer formed by nanofibrils(biopolymer) which are filled inside the deformable material to increase flexibility and decrease the crystalline disorder (Xu et al., 2017). This result in improved stretchability of semiconductors polymer without affecting mobility of charge carrier. Such semiconductor polymers are compared with amorphous silicon structure. Wearable material also fabricated by filling the microchannel network of elastomeric polymer by liquid metal /ionic solution (Kim, et al., 2010; Yamada, 2011; Yoon, Koo and Chang, 2015) has measurable deformation characteristics. This microfluidic technique based conductive fluids become stretchable conductor with super-high stretchability (Zhu et al., 2013). In these liquid filled material piezoresistive property depends on type of liquid fill, cross section structure, depth as well as lateral position of rooted channel. Eutectic gallium indium /galistan gallium based alloy which has nontoxic property and is liquid at room temperature are low cost material. Whereas high price indium limits its large-scale production in practical usage. Ionogel and hydrogel are ionic conductor of elastic solid having potential of design wearable and implantable electronics because of optimal conductivity, stretchability, and transparency.

Hydrogel is biocompatible polymeric networks distended in salty water so they dried out in air through evaporation process. Normally humectants are used to help hydrogels to maintain moisture at low humidity where as polymeric networks distended by ionic solution are known as ionogels but they do not require humectants due to non-volatile nature in vacuum. Traditional conductive material uses electron as charge carrier where as ionic conductor uses ions for response to mechanical signal. With variation in change in shape and size the capacitive and resistive value of material change. This mechanical deformation help in monitoring of pressure of sensor (Sun et al., 2014).

Silicon Nanowire/ Nanostrip

Semiconductors such as nanowires /nanostrips are composition of element/ mixture of Silicon, Zincoxide, Galliumnitride etc., are one dimensional nano geometry with highest degree of mechanical robustness as well as unique electronic/ photonic, electrochemical and thermal properties. This Silicon nanowire piezoresistive effect work as active material in wearable sensors. Silicon is extensively used in design of integrated circuits and RF application in piezoresistive application. This one-dimensional semiconductor unusually has large piezoresistive coefficient in comparison to bulk. It is observed that large piezoresistive effect of nano wires is due to less diameter (approximately 300 nm). It is also noted that extrinsic surface piezoresistance of Silicon nano wires surface charge density carrier durin $SiO2/Al2O3$ passivation (Winkler, Bertagnolli and Lugsein, 2015). These passivated SiNWs has comparable piezo effect with bulk of silicon. Attaching Silicon on surface of soft and deformable polymer substrate results in flexible and stretchable strain gauge, because the applied strain transfer to Silicon via polymer. The network of Silicin nano wire composed of random nanonet containing nanowires in statistical average of individual NW properties. So they have super reproducibility from one device to another device, this unique properties of silicon nanonet is actual device (Yong-Lae, Majidi, Kramer and Berard, 2010; Serre, Mongillo, Periwal, Baron and Teron, 2014).

ZnO Nanowires or Nanobelts

Zinc is an important inorganic material which has piezoelectric property in wurtzite structures (ZnO, GaN, InN, etc.).Zinc oxide nano wires/nanobelts are used to detect electrical signal from mechanical stress. Due to excellent piezoelectric property of zinc oxide it has potential dynamic characteristics of physical variation without extra power supply. The seminal work on zinc oxide were reported to promote practical usage of it (Zang et al., 2015). When nano wire of zinc oxide is combined with flexible polymer it work as strain sensor with zinc metal and form schottky contact between metal. This asymmetric metal (schottky) barrier heights at the ends of electrode change strain and it allow charge to transport through metal-semiconductor(MES) until it not reach to equilibrium. To improve power utilization in practical application individual nano wires /laterally- packaged nano wire are used for flexible electronics. Zinc based oxide is used for large scale wearable application.

Metal Nanowires/ Nanoparticles & Thin Films

Metallic nanowire is used as active element in wearable technology. Gold nanowire were reported where dip-coated tissue paper sandwiched between PDMS sheet as electrode (Gong et al., 2014). On applying pressure loading gold nanowire (AuNW) contacted with electrode pairs as conductive path. Gold nanowire show improved piezoresistive mechanism so it works as flexible sensor with high sensitivity, fast response (in range <17 ms) and high stability. Metal based nanoparticles in two dimensional /three dimensional conductive polymer composites are also reported as piezoresistive sensing elements.

Stretchable Electrode

Stretchable electrode for wearable technology should have high conductivity for large range of stress/strains with reliable performance in repeated and extended stretching task. Metal like silver, gold and copper are very promising nanowires for stretchable electrode due to high conductivity and good mechanical properties. Silver nanowire work as elastic substrate for low temperature nano weldling with high conductivity, transparent and approximately 100% strain holding capacity (Lee et al., 2012). This technique is used for large scale fabrication of material at low cost. Also reported that with improved stability siliver nanowires are used for industrial applications.

Process of Design of Wearable Antenna

In previous section various conductive and substrate materials are discussed for wearable antenna design. So next step is to find electrical properties of these uncharacterized materials like electrical conductivity in case of conductive materials and for substrate material loss tangent and permittivity need to analyzed. After that radiating patch and feeding line geometry of flexible antennas are designed, simulated by EM simulator, and optimized if required than fabricated. Fabrication depends on antenna topology and type of material. Flow chart 2 represent process flow of wearable antenna design.

Fabrication of wearable antenna is based on conductive and substrate materials type. The commercially available techniques like line pattern based on deposition of conductive polymer on surface of substrate is used. In continuation of fabrication technique another way is flexography in which print-making is done on protuberating surface (Board, 2011). Flexography is used in RFID antenna because of high resolution, low production cost, and roll-to-roll production capability. Screen-printing is another method of

Figure 5. Flow Chart: Steps in design of wearable antenna

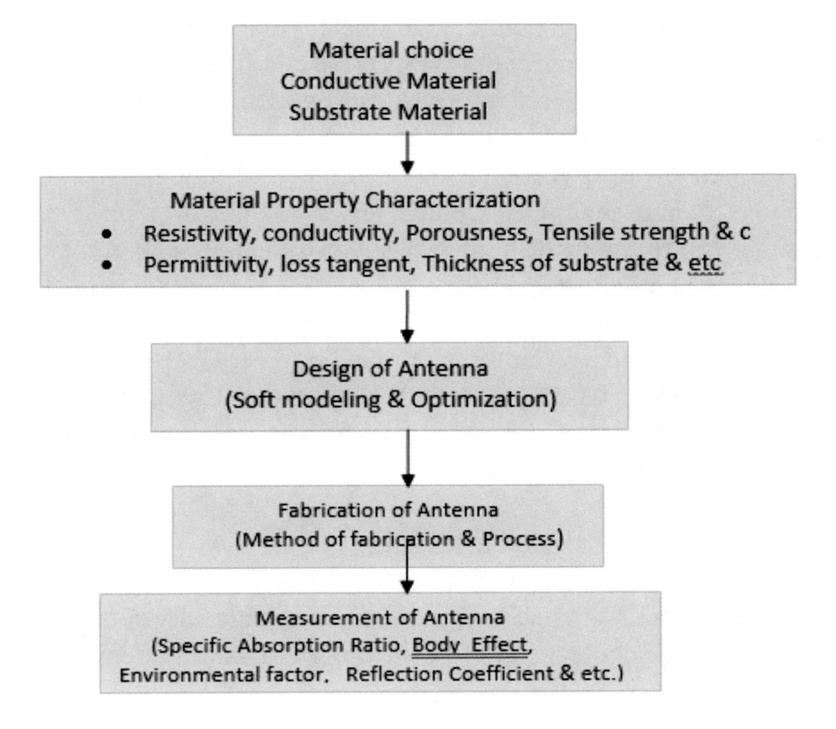

fabrication which is cost-effective for flexible electronics manufacturing. In this technique mask is prepared for desired pattern than developed on flexible surface (Leung and Lam, 2007; Kirsch et al., 2010). Photolithography is most suited technique for PCB design. This technique is also used for fabrication of wearable antenna because of very high accuracy. This technique use photoresist and chemical agents for etching unwanted area metal to get desired pattern. Thermal evaporation used for thin film application where physical vapour deposition technique is used. This method is used in pure material coating film surface (Zhu and Xu, 2012). Sewing and embroidering is based on machine application on textile. In this method in few cases direct adhesion E-textile on surface of fabric but it affect electrical properties of material (Locher, Klemm, Kirstein and Trster, 2006). Inkjet printing technique for fabrication of antenna is done by highly conductive inks. This technique is used for nano-structural materials.

Impedance matching of wearable antennas is one of important factor because flexible substrate material has higher permittivity to achieve maximum miniaturization factors. This increase quality (Q) factor of wearable antenna which results in decrease in bandwidth. So lumped components (capacitors & inductors) are introduced in wearable antenna for tuning. Various technique are utilized for impedance matching of antenna.

In comparison with conventional parameters like S-parameters, radiation pattern and efficiency, more qualitative tests are required for wearable antenna. These measurement include: specific absorption rate, bending impact, thermal test, durability, humidity and robustness.

Specific Absorption Rate measure impact of antenna radiation to human body. Even human body implies negative impact on performance of antennas due to proximity of human tissue. So to analyze parameter return loss impedance bandwidth and radiation patterns simulation experiemts are performed on human body phantom to get realistic result (Hall and Hao, 2006). Permittivity and loss factor in human body vary from low water tissue (Bones) to high water tissue (Muscles). To measure low-power dissipation antenna, vertical input and vertical output measurement technique is a safe and convenient approach.

Parameter of Wearable Material

Bending and crumpling is one of important parameter of wearable antenna. This parameter impact electric field distribution of wearable material which affect performance of antenna (Khaleel, Al-Rizzo and Rucker, 2012). Durability and robustness is another important parameter because it impact functionality and performance of antenna. Return loss and resonant frequency of antenna also shift /deteriorate because of impedance mismatch radiating structures.

Washing Factors (Washability)

Wearable antennas are exposed to sweat, dirt and dust which impact the performance. So wearable antennas integrated in clothing are soaked with water/washed/laundered. That mean consistent performance antenna is required after washing. Washability impact reflection coefficient and radiation parameter of a wearable antenna (Scarpello et al., 2012). This majorly depends on conductive and substrate material.

Environment Factors

Properties of materials are sensitive to environmental variables like humidity and temperature. In some cases wireless systems need to operate in very harsh environments in that case performance stability are essential. In humid condition reflection coefficient and permittivity change from 10% to 90% (Hertleer et al., 2010).

CONCLUSION

This chapter identify the material, fabrication, measurement and important characteristics of wearable antenna.

Various conductive and dielectric substrate materials are discussed in detail. The methodology used for fabrication of wearable antennas for different fabrication material are discussed. The characteristics parameter impedance matching, specific absorption rate, crumpling, bending and washability of wearable antenna are also discussed.

REFERENCES

Allen, M. J., Tung, V. C., & Kaner, R. B. (2010). Honeycomb carbon: A review of grapheme. *Chemical Reviews*, *110*(1), 132–145. doi:10.1021/cr900070d PMID:19610631

Amjadi, M., Yoon, Y. J., & Park, I. (2015). Ultra-stretchable and skin-mountable strain sensors using carbon nanotubes-Ecoflex nanocomposites. *Nanotechnology*, *26*(37), 375–382. doi:10.1088/0957-4484/26/37/375501 PMID:26303117

Baurley, S. (2004). Interactive and experiential design in smart textile products and applications. *Personal and Ubiquitous Computing*, (8), 274–281.

Board, N. (2011). Handbook On Printing Technology (2nd ed.). Academic Press.

Chen, B., Wu, K., & Yao, W. (2004). Conductivity of carbon fiber reinforced cement-based. *Cement and Concrete Composites*, *26*(4), 291–297. doi:10.1016/S0958-9465(02)00138-5

Choong, Shim, Lee, Jeon, Ko, Kang, … Chung. (2014). Silver Nanowires Coated on Cotton for Flexible Pressure Sensors. *Advanced Materials, 26*, 3451–3458.

Coosemans, J., Hermans, B., & Puers, R. (2006). Integrating wireless ECG monitoring in textiles. *Sensors and Actuators. A, Physical*, *130-131*, 48–53. doi:10.1016/j.sna.2005.10.052

Das, N. C., Chaki, T. K., & Khastgir, D. (2002). Effect of axial stretching on electrical resistivity of short carbon fibre and carbon black filled conductive rubber composites. *Polymer International*, *51*(2), 156–163. doi:10.1002/pi.811

Edmison, J., Jones, M., Nakad, Z., & Martin, T. (2002). Using piezoelectric materials for wearable electronic textiles. *Proceedings of the 6th International Symposium on Wearable Computers (ISWC)*, 41–48. 10.1109/ISWC.2002.1167217

Fu, X. W., Liao, Z. M., Zhou, J. X., Zhou, Y. B., Wu, H. C., Zhang, R., ... Yu, D. (2011). The development of graphene-based devices for cell biology research. *Applied Physics Letters*, *99*, 213–217.

Gong, S., Schwalb, W., Wang, Y., Chen, Y., Tang, Y., Si, J., ... Cheng, W. (2014). A wearable and highly sensitive pressure sensor with ultrathin gold nanowires. *Nature Communications*, *5*(1), 1–8. doi:10.1038/ncomms4132 PMID:24495897

Hall, P., & Hao, Y. (2006). Antennas and propagation for body-centric wireless communications. Artech House.

Hammock, M. L., Chortos, A., Tee, B. K., Tok, J. B., & Bao, Z. (2013). The Evolution of Electronic Skin (E-Skin): A brief history, Design Consideration and Recent Progress. *Advanced Materials*, *25*(42), 5997–6038. doi:10.1002/adma.201302240 PMID:24151185

Hempel, M., Nezich, D., Kong, J., & Hofmann, M. (2012). A novel class of strain gauges based on layered percolative films of 2D materials. *Nano Letters*, *12*(11), 5714–5718. doi:10.1021/nl302959a PMID:23045955

Hertleer, C., Van Laere, A., Rogier, H., & Van Langenhove, L. (2010). Influence of relative humidity on textile antenna performance. *Textile Research Journal*, *80*(2), 177–183. doi:10.1177/0040517509105696

Hu, L., Choi, J. W., Yang, Y., Jeong, S., & La, M. F. (2009). Highly conductive paper for energy-storage devices. *Proceedings of the National Academy of Sciences of the United States of America*, *106*(51), 21490–21494. doi:10.1073/pnas.0908858106 PMID:19995965

Hu, L., Hecht, D. S., & Gruner, G. (2010). Carbon Nanotube Thin Films. *Fabrication, Properties, and Applications Chemical Review*, *110*, 5790–5844. PMID:20939616

Hu, L., Yuan, W., Brochu, P., Gruner, G., & Pei, Q. (2009). Highly compliant transparent electrodes. *Applied Physics Letters*, *94*(16), 101–105.

Hu, N., Karube, Y., Yan, C., Masuda, Z., & Fukunaga, H. (2008). Tunneling Effect in a Polymer/Carbon Nanotube Nanocomposite Strain Sensor. *Acta Materialia*, *56*(13), 2929–2936. doi:10.1016/j.actamat.2008.02.030

Huang, G., Xiao, H., & Fu, S. (2015). 2015, Wearable Electronics of Silver-Nanowire/Poly(dimethylsiloxane) Nanocomposite for Smart Clothing. *Scientific Reports*, *5*(1), 13971–13981. doi:10.1038rep13971 PMID:26402056

Huang, M., Pascal, T. A., Kim, H., Goddard, W. A. III, & Greer, J. R. (2011). Electronic--mechanical coupling in graphene from in situ nanoindentation experiments and multiscale atomistic simulations. *Nano Letters*, *11*(3), 1241–1246. doi:10.1021/nl104227t PMID:21309539

Hyun, D. C., Park, M., Park, C., Kim, B., Xia, Y., Hur, J. H., ... Jeong, U. (2011). Ordered Zigzag Stripes of Polymer Gel/Metal Nanoparticle Composites for Highly Stretchable Conductive Electrodes. *Advanced Materials*, *23*(26), 2946–2950. doi:10.1002/adma.201100639 PMID:21590816

Jung, S., Kim, J. H., Kim, J., Choi, S., Lee, J., Park, I., ... Kim, D. (2014). Reverse-Micelle Induced Porous pressure Sensitive Rubber for wearbale for human machine Interfaces. *Advanced Materials*, *26*(28), 4825–4830. doi:10.1002/adma.201401364 PMID:24827418

Jurewicz, Worajittiphon, King, Sellin, Keddie, & Dalton. (2011). Carbon nanotubes buckypaper radiation studies for medical physics applications. *Physics Chemistry B, 115*, 6395–6400.

Khaleel, H. R., Al-Rizzo, H., & Rucker, D. (2012). Compact polyimide based antennas for flexible displays. *IEEE Journal of Display Technology*, *8*(2), 91–97. doi:10.1109/JDT.2011.2164235

Kim, D., Viventi, J., Amsden, J. J., Xiao, J., Vigeland, L., Kim, Y., ... Rogers, J. A. (2010). Dissolvable films of silk fibroin for ultrathin conformal bio-integrated electronics. *Nature Materials*, *9*(6), 511–517. doi:10.1038/nmat2745 PMID:20400953

Kim, D. H., Kim, Y., Wu, J., Liu, Z., Song, J., Kim, H., ... Rogers, J. A. (2009). Ultrathin Silicon Circuits With Strain-Isolation Layers and Mesh Layouts for High-Performance Electronics on Fabric, Vinyl, Leather, and Paper. *Advanced Materials*, *21*(36), 3703–3707. doi:10.1002/adma.200900405

Kirsch, N., Vacirca, N., Kurzweg, T., Fontecchio, A. K., & Dandekar, K. R. (2010). Performance of transparent conductive polymer antennas in a MIMO ad-hoc network. *IEEE 6th International Conference on Wireless and Mobile Computing, Networking and Communications*, 9–14.

Lee, H., Kim, J., Hong, S., & Yoon, J. (2005). Micro-machined CPW-fed suspended patch antenna for 77 GHz automotive radar applications. *3rd European Microwave Conference*, 1707-1710.

Lee, P., Lee, J., Lee, H., Yeo, J., Hong, S., Nam, K. H., ... Ko, S. H. (2012). Highly stretchable and highly conductive metal electrode by very long metal nanowire percolation network. *Advanced Materials*, *24*(25), 3326–3332. doi:10.1002/adma.201200359 PMID:22610599

Lee, S., Reuveny, A., Reeder, J., Lee, S., Jin, H., Liu, Q., ... Someya, T. (2016). A transparent bending-insensitive pressure sensor. *Nature Nanotechnology*, *11*(5), 472–478. doi:10.1038/nnano.2015.324 PMID:26809055

Lee, Y., Bae, S., Jang, H., Jang, S., Zhu, S., Sim, S. H., ... Ahn, J. (2010). Wafer-scale synthesis and transfer of graphene films. *Nano Letters*, *10*(2), 490–493. doi:10.1021/nl903272n PMID:20044841

Leung, S., & Lam, D. (2007). Performance of printed polymer-based RFID antenna on curvilinear surface. *IEEE Transactions on Electronics Packaging Manufacturing*, *30*(3), 200–205. doi:10.1109/TEPM.2007.901181

Li, L., Li, X., Du, M., Guo, Y., Li, Y., Li, H., ... Fang, Y. (2016). n situ formation of a cellular graphene framework in thermoplastic composites leading to superior thermal conductivity. *Journal of Materials Chemistry*, *28*, 3360–3366. doi:10.1021/acs.chemmater.6b00426

Li, X., Yang, T., Yang, Y., Zhu, J., Li, L., Alam, F. E., ... Zhu, H. (2016). Large-Area Ultrathin Graphene Films by Single-Step Marangoni Self-Assembly for Highly Sensitive Strain Sensing, Application. *Advanced Functional Materials*, *26*(9), 1322–1329. doi:10.1002/adfm.201504717

Linz, T., Gourmelon, L., & Langereis, G. (2007). Contactless EMG sensors embroidered onto the textile. *Proceedings of the 4th International Workshop on Wearable and Implantable Body Sensor Networks*, 29–34. 10.1007/978-3-540-70994-7_5

Locher, I., Klemm, M., Kirstein, T., & Trster, G. (2006). Design and characterization of purely textile patch antennas. *IEEE Transactions on Advanced Packaging*, *29*(4), 777–788. doi:10.1109/TADVP.2006.884780

Löfhede, J., Seoane, F. & Thordstein. (2010). Soft textile electrodes for EEG monitoring. *Proceedings of 2010 the 10th IEEE International Conference on Information Technology and Applications in Biomedicine (ITAB)*, *134*, 1–4.

Ma, R., Kang, B., Cho, S., Choi, M., & Baik, S. (2015). Extraordinarily High Conductivity of Stretchable Fibers of Polyurethane and Silver Nanoflowers. *ACS Nano*, *9*(11), 10876–10886. doi:10.1021/acsnano.5b03864 PMID:26485308

Meyer, J., Lukowicz, P., & Tröster, G. (2006). Textile Pressure Sensor for Muscle Activity and Motion Detection. *Proceeding of the 10th IEEE International Symposium on Wearable Computers*, 11–14. 10.1109/ISWC.2006.286346

Nathan, A., Ahnood, A., Cole, M. T., Lee, S., Suzuki, Y., Hiralal, P., ... Milne, W. I. (2012). Flexible Electronics: The Next Ubiquitous Platform. *Proceedings of the IEEE*, *100*, 1486–1517. doi:10.1109/JPROC.2012.2190168

Novoselov, K. S., Fal, V. I., Colombo, L., Gellert, P. R., Schwab, M. G., & Kim, K. (2012). A roadmap for graphene. *Nature*, *490*, 192–200. doi:10.1038/nature11458 PMID:23060189

Obitayo, Liu, & Sens. (2012). A Review: Carbon Nanotube-Based Piezoresistive Strain Sensors. *Journal of Sensor*, 1–15.

Pan, L., Chortos, A., Yu, G., Wang, Y., Isaacson, S., Allen, R., ... Bao, Z. (2014). An ultra-sensitive resistive pressure sensor based on hollow-sphere microstructure induced elasticity in conducting polymer film. *Nature Communications*, *5*(1), 300–312. doi:10.1038/ncomms4002 PMID:24389734

Pereira, V. M., Neto, A. C., & Peres, N. (2009). Tight-binding approach to uniaxial strain in grapheme. *Physical Review B: Condensed Matter and Materials Physics*, *80*(4), 454–460. doi:10.1103/PhysRevB.80.045401

Rim, Y., Bae, S., Chen, H., De Marco, N., & Yang, Y. (2016). Recent progress in Material and Devices towards Printable and Flexible sensors. *Advanced Materials*, *28*(22), 4415–4440. doi:10.1002/adma.201505118 PMID:26898945

Russo, A., Ahn, B. Y., Adams, J. J., Duoss, E. B., Bernhard, J. T., & Lewis, J. A. (2011). Pen-on-Paper Flexible Electronics. *Advanced Materials*, *23*(30), 3426–3430. doi:10.1002/adma.201101328 PMID:21688330

Scarpello, M., Kazani, I., Hertleer, C., Rogier, H., & Ginste, D. (2012). Stability and efficiency of screen-printed wearable and washable antennas. *IEEE Antennas and Wireless Propagation Letters*, *11*, 838–841. doi:10.1109/LAWP.2012.2207941

Serre, P., Mongillo, M., Periwal, P., Baron, T., & Ternon, C. (2014). Percolating silicon. *Nanotechnology*, *26*(1), 15201. doi:10.1088/0957-4484/26/1/015201 PMID:25483713

Stankovich, S., Dikin, D. A., Dommett, G. H., Kohlhaas, K. M., Zimney, E. J., Stach, E. A., ... Ruoff, R. S. (2006). Graphene-based composite materials. *Nature*, *44*(2), 282–286. doi:10.1038/nature04969 PMID:16855586

Sun, J., Keplinger, C., Whitesides, G. M., & Suo, Z. (2014). Ionic skin. *Advanced Materials*, *26*(45), 7608–7614. doi:10.1002/adma.201403441 PMID:25355528

Tao, H., Hwang, S., Marelli, B., An, B., Moreau, J. E., Yang, M., ... Omenetto, F. G. (2014). Silk-based resorbable electronic devices for remotely controlled therapy and in vivo infection abatement. *Proceedings of the National Academy of Sciences of the United States of America*, *111*(49), 17385–17389. doi:10.1073/pnas.1407743111 PMID:25422476

Wang, Y., Wang, L., Yang, T., Li, X., Zang, X., Zhu, M., ... Zhu, H. (2014). Wearable and Highly Sensitive Graphene Strain Sensors for Human Motion Monitoring. *Advanced Functional Materials*, *24*(29), 4666–4670. doi:10.1002/adfm.201400379

Wang, Y., Yang, T., Lao, J., Zhang, R., Zhang, Y., Zhu, M., ... Yu, W. (2015). Ultra-sensitive graphene strain sensor for sound signal acquisition and recognition. *Nano Research*, *8*(5), 1627–1636. doi:10.100712274-014-0652-3

Winkler, K., Bertagnolli, E., & Lugstein, A. (2015). Origin of Anomalous Piezoresistive Effects in VLS Grown Si Nanowires. *Nano Letters*, *15*(3), 1780–1785. doi:10.1021/nl5044743 PMID:25651106

Xu, J., Wang, S., Wang, G. N., Zhu, C., Luo, S., Jin, L., ... To, J. W. (2017). Highly stretchable polymer semiconductor films through the nanoconfinement effect. *Science*, *355*(6320), 59–64. doi:10.1126cience.aah4496 PMID:28059762

Yamada, T., Hayamizu, Y., Yamamoto, Y., Yomogida, Y., Izadi-Najafabadi, A., Futaba, D. N., & Hata, K. (2011). A stretchable carbon nanotube strain sensor for human-motion detection. *Nature Nanotechnology*, *6*(5), 296–301. doi:10.1038/nnano.2011.36 PMID:21441912

Yang, T., Wang, W., Zhang, H., Li, X., Shi, J., He, Y., ... Zhu, H. (2015). Tactile Sensing System Based on Arrays of Graphene Woven Microfabrics: Electromechanical Behavior and Electronic Skin Application. *ACS Nano*, *9*(11), 10867–10875. doi:10.1021/acsnano.5b03851 PMID:26468735

Yang, T., Wang, Y., Li, X., Zhang, Y., Li, X., Wang, K., ... Zhu, H. (2014). Torsion sensors of high sensitivity and wide dynamic range based on a graphene woven structure. *Nanoscale*, *6*(21), 13053–13059. doi:10.1039/C4NR03252G PMID:25247375

Yao, H., Ge, J., Wang, C., Wang, X., Hu, W., Zheng, Z., ... Yu, S. (2013). A flexible and highly pressure-sensitive graphene-polyurethane sponge based on fractured microstructure design. *Advanced Materials*, *25*(46), 6692–6698. doi:10.1002/adma.201303041 PMID:24027108

Yong-Lae, Majidi, Kramer, & Berard. (2010). *Hyperelastic Pressure Sensing with a Liquid Embedded Elastomer. Journal of Micromechanics and Microengineering*, *20* (12), 125026–125029.

Yoon, S. G., Koo, H., & Chang, S. T. (2015). Highly stretchable and transparent microfluidic strain sensors. *Applied Material Interfaces*, *7*(49), 27562–27570. doi:10.1021/acsami.5b08404 PMID:26588166

Zang, Y., Zhang, F., Huang, D., Gao, X., Di, C., & Zhu, D. (2015). Flexible suspended gate organic thin-film transistors for ultra-sensitive pressure detection. *Nature Communications*, *6*(1), 6269–6277. doi:10.1038/ncomms7269 PMID:25872157

Zheng, Y., He, Z., Gao, Y., & Liu, J. (2013). 2013, Direct Desktop Printed-Circuits-on-Paper Flexible Electronics. *Scientific Reports*, *3*(1), 1786–1791. doi:10.1038rep01786

Zhu, B., Wang, H., Leow, W. R., Cai, Y., Loh, X. J., Han, M., & Chen, X. (2016). Silk Fibroin for Flexible Electronic Devices. *Advanced Materials*, *28*(22), 4250–4265. doi:10.1002/adma.201504276 PMID:26684370

Zhu, S., So, J. H., Mays, R., Desai, S., Barnes, W. R., Pourdeyhimi, B., & Dickey, M. D. (2013). Ultrastretchable Fibers with Metallic Conductivity Using a Liquid Metal Alloy Core. *Advanced Functional Materials*, *23*(18), 2308–2314. doi:10.1002/adfm.201202405

Zhu, Y., & Xu, F. (2012). Buckling of aligned carbon nanotubes as stretchable conductors: A new manufacturing strategy. *Advanced Materials*, *24*(8), 1073–1077. doi:10.1002/adma.201103382 PMID:22271642

Chapter 13
Relationship Between Co-Axial Probe Feed and Inset Feed in Rectangular Microstrip Patch

Ravi Kant Prasad
Bundelkhand Institute of Engineering and Technology Jhansi, India

D. K. Srivastava
Bundelkhand Institute of Engineering and Technology Jhansi, India

Rishabh Kumar Baudh
https://orcid.org/0000-0003-1026-2373
Bundelkhand Institute of Engineering and Technology Jhansi, India

J. P. Saini
Bundelkhand Institute of Engineering and Technology Jhansi, India

ABSTRACT

In this chapter, an equation is obtained using curve fit formula that shows the relationship between the simulated co-axial probe feed distance and theoretically inset feed distance in rectangular microstrip patch antenna. The simulation process is performed using IE3D simulation software tool and theoretical calculation performed by the cavity model. Using this equation, one can avoid hit and trial for getting simulated co-axial feed distance by knowing theoretically inset feed distance. A ratio also has been developed between co-axial probe feed and inset probe feed.

DOI: 10.4018/978-1-5225-9683-7.ch013

INTRODUCTION

Modern communication system widely used microstrip patch antenna because of its lightweight, compact and cost effective constraints. In the design of patch antenna, the feeding method plays a significant role. The input impedance of patch antenna depends on the feeding type and theirposition. By varying the feed position an analysis of the impedance variation has been performed in microstrip line fed patch antenna. For this substrate's size and finite metallization thickness with various lengths of the feeding line as well as losses are taking accounts (Snezana et. al, 2014). The contacting scheme such as co-axial probe feed and microstrip line, have RF power directly to the radiating patch and electromagnetic field coupling has been used by other contacting scheme such as proximity and aperture coupled feed, for the transferring power between the microstrip line and radiating patch (Varshney et. al, 2014). Microstrip patch antenna has so many methods for the analysis, in which cavity, transmission line and full wave methods are more popular methods. Based on transmission line method a microstrip patch antenna has been designed for 2.45 GHz (Barrou et. al, n.d.).

A transmission line model has been used for analysis and a curve fit formula has been presented for locating the exact inset feed to obtained 50Ω input impedance with microstrip line inset feed (Ramesh and Yip, 2003). Full wave analysis, transmission line and cavity model has been used for analytical study of input impedance behavior of co-axial probe feed patch antenna (Pozar, 1982 & Garg, Bhartia, and Ittipiboon, 2001). The input impedance of co-axial probe feed and inset feed patch antenna exhibit $\cos^2 (\pi Y_0/L)$ and $\cos^4 (\pi Y_0/L)$ behavior respectively with Y_0 is the feed point position from the radiating edge along the direction of length (L) (Basilio et. al, 2001).

A theoretical investigation has been performed which shows that the input impedance dependent on feed position of rectangular patch antenna (Ghatak and Pal, 2015 & Samaras, Stefanovski, and Branko, 2014). The input impedance can be adjusted by varying the feed point location. To obtain a good matching between the generated impedance and input impedance, it has been required to determination of exact feed point.

In this work, an exact inset feed distance is determined using model expansion analysis method for good impedance matching and verified by simulation process. In addition, the coaxial probe feed distance is determined by simulation process and theoretically verified using cavity model analysis method. A relationship is developed between the theoretical inset feed distance and simulated co-axial probe feed distance using curve fitting formula. Zeeland IE3D version 9.0, Graphmatica and MATLAB are used as simulation, curve fitting and theoretical analysis respectively.

MEASUREMENT TECHNIQUES

In rectangular patch antenna the co-axial probe feed and the microstrip line inset feed has been used for impedance matching and measurements. The glass epoxy FR4 substrate having thickness, dielectric constant and loss tangent of 1.6 mm, 4.4 and 0.01respectively has been used.

The range of frequency for which relationship between both feed positions has been developed lies between 2 to 6 GHz. The rectangular patch length and width, for particular frequency can be calculated. The length and width are the two primary parameters that are needed in design of the rectangular patch antenna. The width (W) of rectangular patch antenna is calculated as follows (Constantine and Balanis, 2005)

$$W = \frac{0.5C}{f}\sqrt{\frac{2}{\varepsilon_r + 1}}$$

Where C = Light's speed (3×10^{11} mm/s), f = Antenna design frequency, ε_r = Dielectric constant of substrate

The effective dielectric constant ε_{re} is calculated as

$$\varepsilon_{re} = \frac{\varepsilon_r + 1}{2} + \frac{\varepsilon_r - 1}{2}\left[1 + 12\frac{h}{W}\right]^{-0.5}$$

The extended patch length ΔL is calculated as

$$\Delta L = 0.412h\frac{(\varepsilon_{re} + 0.3)(W + 0.26h)}{(\varepsilon_{re} - 0.258)(W + 0.8h)}$$

and actual value of the patch length is calculated as follows

$$L = \frac{0.5C}{f\sqrt{\varepsilon_{re}}} - 2\Delta L$$

Figure 1a. Co-axial probe feed rectangular patch antenna

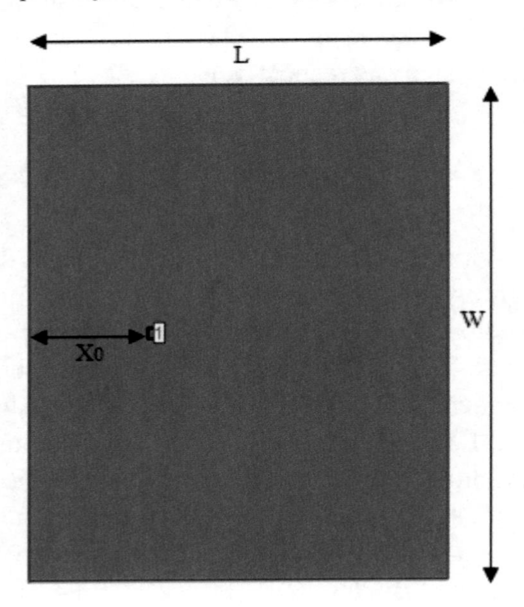

Figure 1b. Microstrip line feed rectangular patch antenna

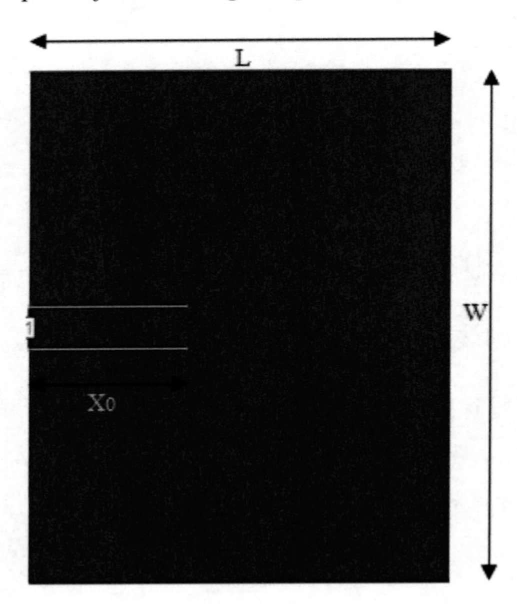

Figure 2. Simulated return loss for co-axial probe feed (at 7.06mm)

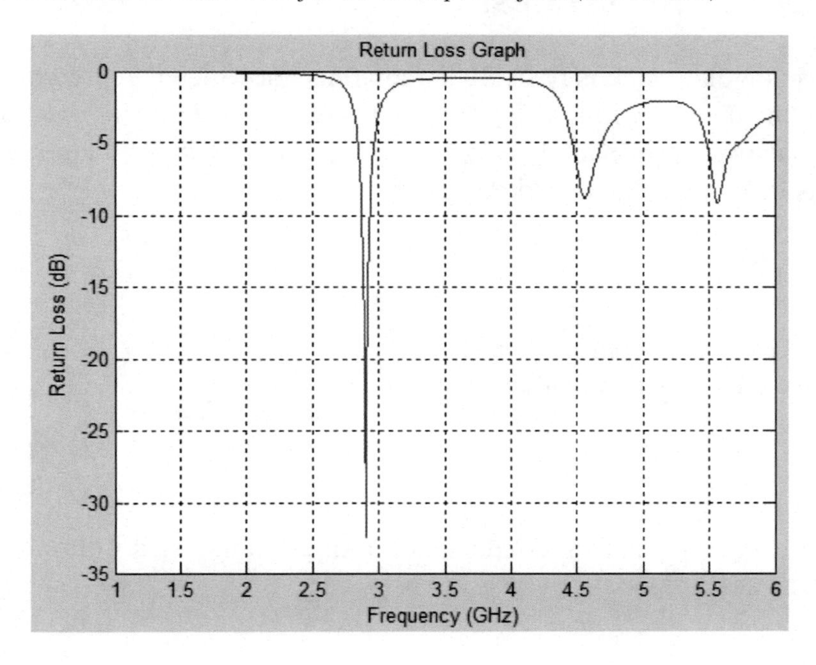

Figure 3. Simulated return loss for inset feed (at 8.6884mm)

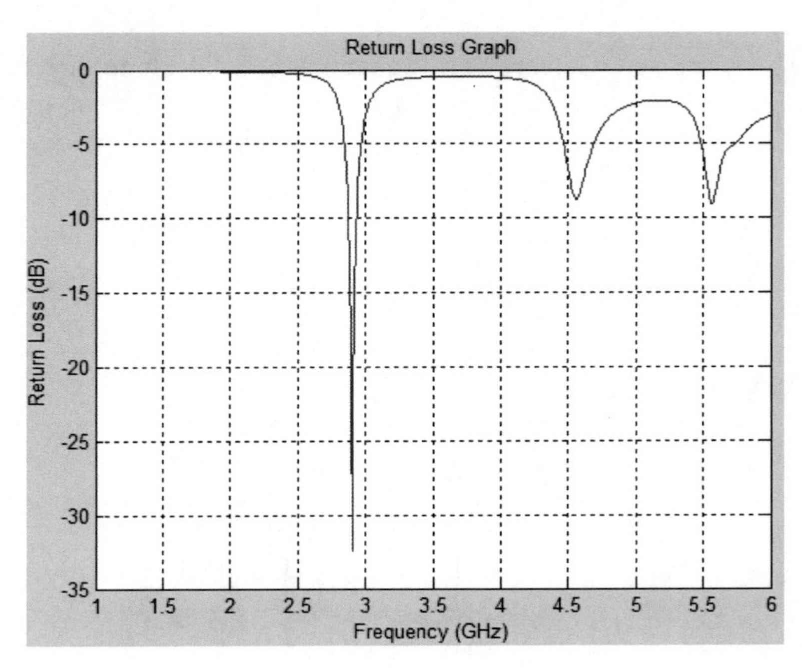

THEORETICAL ANALYSIS

The rectangular patch is basically a radiating slot which is represented by a parallel equivalent admittance(Y) which is shown in figure 4. The radiating patch has two edges namely edge1 and edge2.Theequivalent admittance of edge is given as (Ghatak and Pal, 2015)

$$Y = G + jB$$

Where G = Conductance, B = Susceptance

Since edge 2 is identical to edge1, thus

$$Y = Y_1 = Y_2, G = G_1 = G_2$$

In rectangular patch the conductance of single radiating slot of finite width W by the cavity model can be given as

$$G = \frac{W}{120\lambda_0}\left[1 - \frac{1}{24}(k_0 h)^2\right]$$

Using admittance transformation equation, the total admittance at edge 1 can be calculated by transferring the admittance of edge 2 from the output to input terminals. At edge 2, the transformed admittance is given as

$$Y_2 = G_2 + jB_2 = G_1 - jB_1$$

or

$$G_2 = G_1,\ B_2 = -B_1$$

Figure 4. Equivalent admittance circuit of rectangular patch

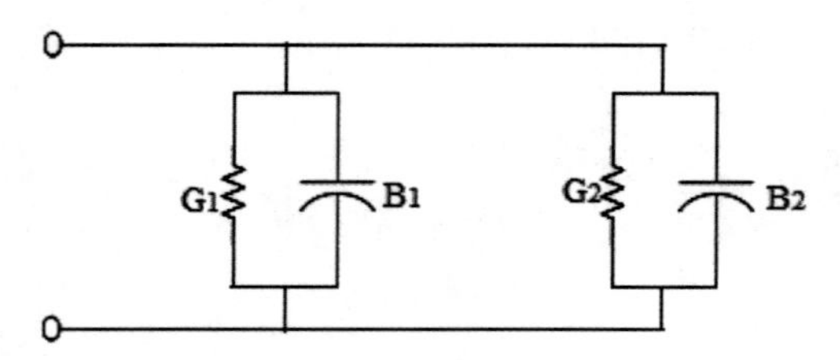

Therefore, the total real resonant input admittance is given as

$$Y_{in} = Y_1 + Y_2 = 2G_1$$

Now, the real resonant input impedance, without taking into account mutual effects between the edges is given as

$$Z_{in} = \frac{1}{Y_{in}} = R_{in} = \frac{1}{2G_1}$$

The resonant input impedance after taking into account mutual effects between the edges is given as

$$R_{in} = \frac{1}{2\left(G_1 + G_{12}\right)}$$

The mutual conductance G_{12} can be calculated as

$$G_{12} = \frac{1}{120\pi^2}\int_0^{\pi}\left[\frac{\sin\left(\frac{k_0 W}{2}\cos\theta\right)}{\cos\theta}\right]^2 J_0\left(k_0 L\sin\theta\right)\sin^3\theta d\theta$$

Where, J_0 is the first kind zero order Bessel's function. The mutual conductance G_{12} is small in comparison to the self conductance G_1.

The resonant input impedance $R_{in}(x_0=0)$, is referenced at edge1. It can be changed by using an inset feed, recessed a distance x_0 from edge1, as shown in figure 1. Using model expansion analysis, the input impedance for the inset feed is calculated as

$$R_{in}\left(x = x_0\right) = \frac{1}{2\left(G_1 + G_{12}\right)}\cos^2\left(\frac{2\pi}{L}x_0\right) = R_{in}\left(x = 0\right)\cos^2\left(\frac{\pi}{L}x_0\right)$$

The input impedance is maximum that is 150-300 Ω at the edge ($x_0=0$) and decreases monotonically as the inset feed point moves from the edge towards the center of the patch. The changes in input impedance takes place by varying the feed point location with the functioncos² $(\pi x_0/L)$.Since the input impedance is maximum at the radiating edge and desired impedance is 50Ω, thus the inset feed point distance x_0 is calculated as

$$x_0 = \left(\frac{L}{\pi}\right)\cos^{-1}\left(\left(\frac{50}{R_{in}\left(x=0\right)}\right)^{0.5}\right)$$

Figure 5. Theoretical return loss for inset feed (at 8.6884 mm)

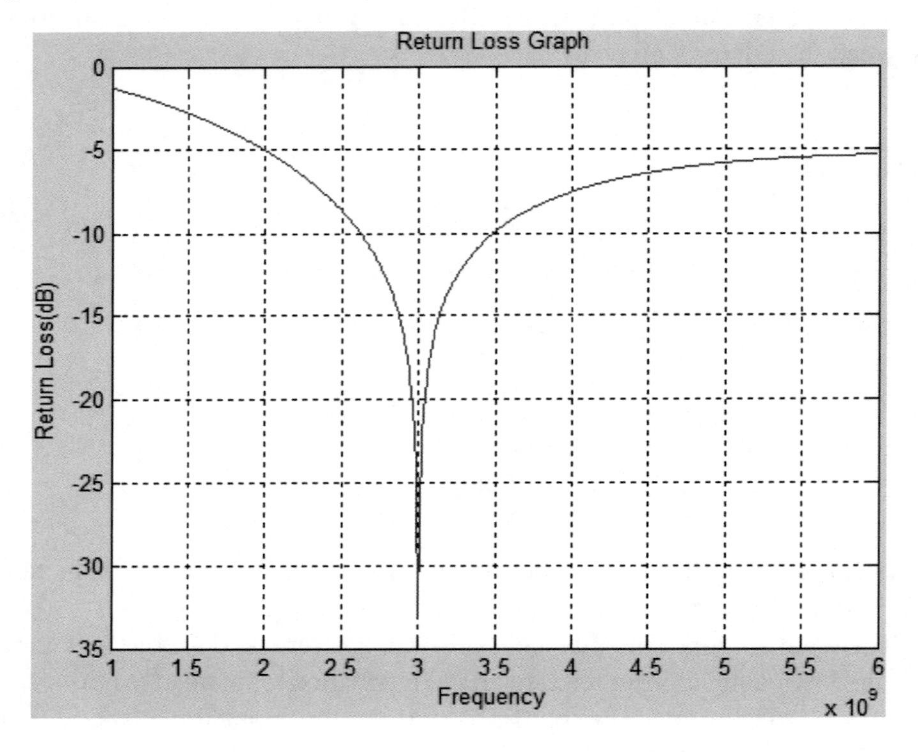

Figure 6. Equivalent circuit of simple rectangular patch

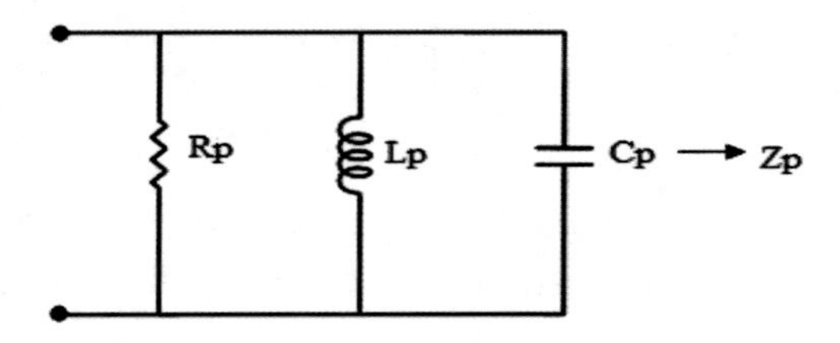

The microstrip patch antenna is narrow-band resonant antenna and termed as lossy cavities. Therefore, for the analysis of input impedance of patch antenna cavity model becomes a natural choice. According to this model the simple rectangular patch is analyzed as a parallel arrangement of Rp (resistance), Cp (capacitance) and Lp (inductance) as shown in figure 6.

The values of lumped elements Rp, Lp and Cp can be calculated as

$$R_p = \frac{2h\mu_0 C^2}{2\pi f \varepsilon_r \delta_{eff} L W} cos^2\left(\frac{\pi x_0}{L}\right)$$

$$L_p = \frac{R_p \delta_{eff}}{2\pi f}$$

$$C_p = \frac{1}{2\pi f R_p \delta_{eff}}$$

Now the resonant input impedance can be given as

$$Z_{in} = R_{in} + jX_{in} = \frac{1}{\frac{1}{R} + j\omega C + \frac{1}{j\omega L}} = Z_p = \frac{1}{\sqrt{\left(\frac{1}{R_p}\right)^2 + \left(\omega C_p - \frac{1}{\omega L_p}\right)^2}}$$

Where, h= Substrate thickness, μ_o = Substrate permeability, ε_r= Relative permittivity of the substrate, x_o = Feed point on x-axis, δ_{eff} = Effective loss tangent

Using the above values, VSWR, reflection co-efficient and return loss of simple patch can be calculated as

$$VSWR = \frac{1+|\Gamma|}{1-|\Gamma|}$$

$$Reflection\ coefficient(\Gamma) = \frac{Z_p - Z_0}{Z_p + Z_0}$$

Where, Z_0=Feed characteristics impedance (50 Ω) and $ReturnLoss = 10\log|\Gamma|$

Figure 7. Theoretical return loss for co-axial probe feed (at 7.06 mm)

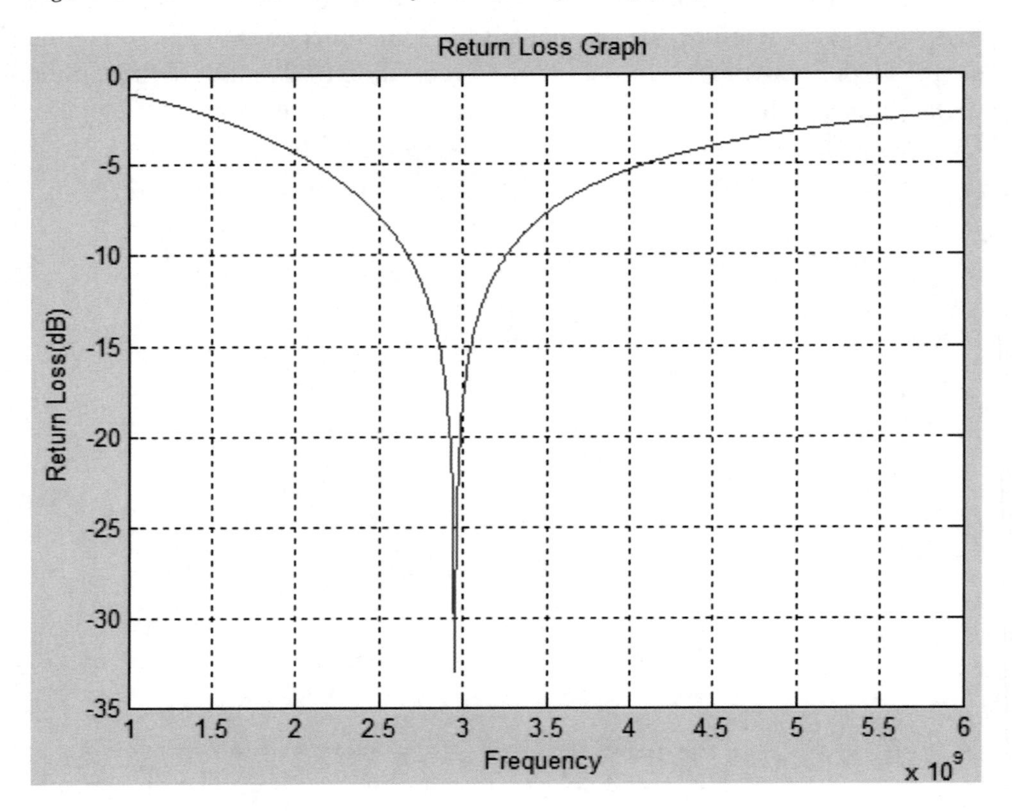

RESULTS AND DISCUSSION

For the frequency range 2GHz to 6GHz, the feed point distance x_0 has been calculated theoretically for the microstrip line feed rectangular patch antenna and simulatically for the co-axial probe feed rectangular patch antenna. The geometry of co-axial probe feed and microstrip line feed rectangular patch antenna with feed point distance x_0 is shown in figure 1 a and figure 1 b respectively. The frequency interval 0.2 GHz has been taken. In simulation process, finding the probe feed distance for impedance matching is tedious work. For this hit and trial method has been used.

But calculation of inset feed distance for proper impedance matching is theoretically easy. Using this theoretically calculated inset feed distance in simulation process, it has been achieved the proper impedance matching and it has been observed that with proper impedance matching, the simulated and

Table 1. Variables and results of rectangular patch antenna

Design Frequency (GHz)	Length (mm)	Width (mm)	Theoretical Inset Feed Distance (mm) (I)	Simulated Co-axial ProbeFeed Distance (mm) (Co)	Ratio (I/Co)
2	35.44	45.64	13.1506	10.33	1.273
2.2	32.17	41.49	11.9356	9.5	1.256
2.4	29.44	38.03	10.9221	8.76	1.247
2.6	27.13	35.11	10.0638	8.12	1.239
2.8	25.15	32.6	9.3273	7.56	1.234
3	23.43	30.43	8.6884	7.06	1.231
3.2	21.93	28.53	8.1288	6.64	1.224
3.4	20.6	26.85	7.6346	6.26	1.220
3.6	19.41	25.36	7.1949	5.9	1.219
3.8	18.35	24.02	6.8011	5.58	1.219
4	17.4	22.82	6.4463	5.3	1.216
4.2	16.53	21.74	6.1251	5.04	1.215
4.4	15.75	20.75	5.8327	4.81	1.213
4.6	15.03	19.85	5.5656	4.59	1.213
4.8	14.37	19.02	5.3205	4.39	1.212
5	13.76	18.26	5.0948	4.21	1.210
5.2	13.2	17.56	4.8863	4.04	1.209
5.4	12.68	16.9	4.6931	3.885	1.208
5.6	12.2	16.3	4.5135	3.737	1.208
5.8	11.75	15.74	4.3462	3.6	1.207
6	11.33	15.21	4.19	3.478	1.205

Table 2. Comparison of resonance peak

Design Frequency	Theoretical Resonance Peak (Probe Feed at 7.06 mm)	Simulated Resonance Peak (Probe Feed at 7.06 mm)	Theoretical Resonance Peak (Inset Feed at 8.6884 mm)	Simulated resonance Peak (Inset Feed at 8.6884 mm)
3 GHz	2.952 GHz	2.902GHz	3.002GHz	2.972 GHz

theoretical return loss graph shows almost equal resonance frequency as the design frequency of patch antenna.

In simulation process, a hit and trial method has been used for getting the co-axial probe feed distance at which proper impedance matching achieved and it gives the approx resonant frequency as design frequency. But when this simulated probe feed distance has been used in theoretical calculation for getting return loss, it gives almost equal resonance frequency as the design frequency of patch antenna.

The design frequency with their calculated length and width, theoretically calculated inset feed distance, simulated probe feed distance using hit and trial for proper impedance matching and ratio of inset feed distance and co-axial feed distance are shown in table 1 as variables and results of rectangular patch antenna.

For 3 GHz design frequency, the simulated co-axial probe feed distance is 7.06 mm at which proper impedance matching has been achieved. The simulated return loss and theoretical return loss for this simulated coaxial probe feed distance is shown in figure 2 and figure 7 respectively.

For 3 GHz design frequency, the theoretical inset feed distance is 8.6884 at which proper impedance matching has been achieved. The simulated return loss and theoretical return loss for this inset feed distance is shown in figure 3 and figure 5 respectively.

From all figures of return loss, it has been observed that the design frequency and the resonance peak frequency in both simulated and theoretical graphs are approximately same, which is also shown in above table 2 as comparison of resonance peak.

Applying the values of simulated co-axial probe feed distance along y-axis and theoretical inset feed distance along x-axis from table 1, in graphmatica (curve fitting software), a relationship has been developed in form of equation and given as

$$y = 5.4214 \times 10^{-6} \times x^4 - 0.0013 \times x^3 + 0.0249 \times x^2 + 0.6424 \times x + 0.4651$$

Knowing theoretical inset feed distance and using above equation, it can be easily determine the co-axial probe feed distance without doing hit and trial in simulation process for getting proper impedance matching or getting maximum negative return loss.

For an example putting the theoretical inset feed distance of 4 GHz design frequency as parameter x=6.4463 mm in above derive equation, it has been obtained co-axial feed distance as parameter y=5.3020 mm. Using this co-axial feed distance in simulation process and theoretical calculation of probe feed, the resonance frequency with return loss is 3.843 GHz,-54.14 dB and 3.953 GHz,-36.157 dB respectively. Thus it can be easily find the co-axial probe feed point in simulation process for impedance matching and getting resonance frequency with return loss.

A ratio also has been developed between theoretical inset feed distance and simulated co-axial probe feed distance. From Table1, it is also observed that the ratio between inset feed distance and co-axial probe feed distance decreases as design frequency increases from lower frequency range to higher frequency range. The range of design frequency is from 2 GHz to 6 GHz. The mean of the ratio is 1.217.

CONCLUSION

A relationship is developed between the theoretical inset feed distance and simulated coaxial feed distance of rectangular patch antenna. A ratio also has been developed between the inset feed distance and coaxial feed distance. By knowing one feed distance other feed distance can be easily determined. This relationship can help in simulation process of design a co-axial probe feed rectangular patch antenna.

REFERENCES

Barrou, O., El Amri, A., & Reha, A. (n.d.). Comparison of Feeding Modes for a Rectangular Microstrip Patch Antenna for 2.45 GHz Applications. *Advances in Ubiquitous Networking, 2*, 457-469.

Basilio, L. I., Khayat, M. A., Williams, J. T., & Long, S. A. (2001). The Dependence of the Input Impedance on Feed Position of Probe and Microstrip Line-fed Patch Antennas. *IEEE Transactions on Antennas and Propagation, 49*(1), 45–47. doi:10.1109/8.910528

Constantine, G., & Balanis, A. (2005). *Antenna Theory, Analysis and Design*. Hoboken, NJ: Wiley.

Garg, R., Bhartia, P., Bahl, I., & Ittipiboon, A. (2001). *Microstrip Antenna Design Handbook*. Norwood, MA: Artech House.

Ghatak, R., & Pal, M. (2015). Revisiting Relations for Modeling the Input Resistance of a Rectangular Microstrip Antenna. *IEEE Antennas & Propagation Magazine, 57*(4), 116–119. doi:10.1109/MAP.2015.2453887

Pozar, D. M. (1982). Input impedance and mutual coupling of rectangular microstrip antennas. *IEEE Transactions on Antennas and Propagation, 30*(6), 1191–1196. doi:10.1109/TAP.1982.1142934

Ramesh, M., & Yip, K. B. (2003). Design Formula for Inset Fed Microstrip Patch Antenna. *Journal of Microwaves and Optoelectronics., 3*(3), 5–10.

Samaras, T., Kouloglou, A., & Sahalos, J. N. (2004). A note on the impedance variation with feed position of a rectangular microstrip antenna. *IEEE Antennas & Propagation Magazine, 46*(2), 90–92. doi:10.1109/MAP.2004.1305543

Snezana, L. (2014). The Impedance Variation with Feed Position of a Microstrip Line-Fed Patch Antenna. *SerbianJournal of Electrical Engineering, 11*(1), 85–96. doi:10.2298/SJEE131121008S

Varshney, H.K., Kumar, M., & Jaiswal, A.K., Saxena, R., & Jaiswal, K. (2014). A Survey on Different Feeding Techniques of Rectangular Microstrip Patch Antenna. *International Journal of Current Engineering and Technology., 4*(3), 1418–1423.

Related Readings

To continue IGI Global's long-standing tradition of advancing innovation through emerging research, please find below a compiled list of recommended IGI Global book chapters and journal articles in the areas of wearable technology, data transmission, and medical applications. These related readings will provide additional information and guidance to further enrich your knowledge and assist you with your own research.

Abbas, R., Michael, K., & Michael, M. G. (2017). What Can People Do with Your Spatial Data?: Socio-Ethical Scenarios. In A. Marrington, D. Kerr, & J. Gammack (Eds.), *Managing Security Issues and the Hidden Dangers of Wearable Technologies* (pp. 206–237). Hershey, PA: IGI Global. doi:10.4018/978-1-5225-1016-1.ch009

Afyf, A., Bellarbi, L., Latrach, M., Gaviot, E., Camberlein, L., Sennouni, M. A., & Yaakoui, N. (2018). Wearable Antennas: Breast Cancer Detection. In S. Delabrida Silva, R. Rabelo Oliveira, & A. Loureiro (Eds.), *Examining Developments and Applications of Wearable Devices in Modern Society* (pp. 161–202). Hershey, PA: IGI Global. doi:10.4018/978-1-5225-3290-3.ch007

Amorim, V. J., Delabrida Silva, S. E., & Oliveira, R. A. (2018). Wearables Operating Systems: A Comparison Based on Relevant Constraints. In S. Delabrida Silva, R. Rabelo Oliveira, & A. Loureiro (Eds.), *Examining Developments and Applications of Wearable Devices in Modern Society* (pp. 86–106). Hershey, PA: IGI Global. doi:10.4018/978-1-5225-3290-3.ch004

Amyx, S. (2017). Privacy Dangers of Wearables and the Internet of Things. In A. Marrington, D. Kerr, & J. Gammack (Eds.), *Managing Security Issues and the Hidden Dangers of Wearable Technologies* (pp. 131–160). Hershey, PA: IGI Global. doi:10.4018/978-1-5225-1016-1.ch006

Angelova, R. A. (2018). Wearable Technologies for Helping Human Thermophysiological Comfort. In S. Delabrida Silva, R. Rabelo Oliveira, & A. Loureiro (Eds.), *Examining Developments and Applications of Wearable Devices in Modern Society* (pp. 203–231). Hershey, PA: IGI Global. doi:10.4018/978-1-5225-3290-3.ch008

Antero, M. C. (2017). Model Course Syllabus: Management of Security Issues in Wearable Technology. In A. Marrington, D. Kerr, & J. Gammack (Eds.), *Managing Security Issues and the Hidden Dangers of Wearable Technologies* (pp. 267–288). Hershey, PA: IGI Global. doi:10.4018/978-1-5225-1016-1.ch011

Applin, S. A., & Fischer, M. D. (2017). Thing Theory: Connecting Humans to Smart Healthcare. In C. Reis & M. Maximiano (Eds.), *Internet of Things and Advanced Application in Healthcare* (pp. 249–265). Hershey, PA: IGI Global. doi:10.4018/978-1-5225-1820-4.ch009

Archondakis, S., Vavoulidis, E., Nasioutziki, M., Oustampasidou, O., Daniilidis, A., Vatopoulou, A., ... Dinas, K. (2019). Mobile Health Applications and Cloud Computing in Cytopathology: Benefits and Potential. In A. Moumtzoglou (Ed.), *Mobile Health Applications for Quality Healthcare Delivery* (pp. 165–202). Hershey, PA: IGI Global. doi:10.4018/978-1-5225-8021-8.ch008

Baihan, M. S., Sánchez, Y. K., Shao, X., Gilman, C., Demurjian, S. A., & Agresta, T. P. (2018). A Blueprint for Designing and Developing M-Health Applications for Diverse Stakeholders Utilizing FHIR. In R. Rajkumar (Ed.), *Contemporary Applications of Mobile Computing in Healthcare Settings* (pp. 85–124). Hershey, PA: IGI Global. doi:10.4018/978-1-5225-5036-5.ch006

Berrahal, S., & Boudriga, N. (2017). The Risks of Wearable Technologies to Individuals and Organizations. In A. Marrington, D. Kerr, & J. Gammack (Eds.), *Managing Security Issues and the Hidden Dangers of Wearable Technologies* (pp. 18–46). Hershey, PA: IGI Global. doi:10.4018/978-1-5225-1016-1.ch002

Black, I., & White, G. (2017). Citizen Science, Air Quality, and the Internet of Things. In C. Reis & M. Maximiano (Eds.), *Internet of Things and Advanced Application in Healthcare* (pp. 138–169). Hershey, PA: IGI Global. doi:10.4018/978-1-5225-1820-4.ch005

Boubeta-Puig, J., Ortiz, G., & Medina-Bulo, I. (2017). Preventing Health Risks Caused by Unhealthy Air Quality Using a CEP-Based SOA 2.0. In C. Reis & M. Maximiano (Eds.), *Internet of Things and Advanced Application in Healthcare* (pp. 170–196). Hershey, PA: IGI Global. doi:10.4018/978-1-5225-1820-4.ch006

Bouchemal, N., Maamri, R., & Bouchemal, N. (2019). Telemonitoring Healthcare System-Based Mobile Agent Technology. In N. Bouchemal (Ed.), *Intelligent Systems for Healthcare Management and Delivery* (pp. 198–205). Hershey, PA: IGI Global. doi:10.4018/978-1-5225-7071-4.ch008

Boussebough, I., Chaib, I. E., & Boudjit, B. (2019). An Ambient Multi-Agent System for Healthcare Monitoring of Patients With Chronic Diseases. In N. Bouchemal (Ed.), *Intelligent Systems for Healthcare Management and Delivery* (pp. 61–71). Hershey, PA: IGI Global. doi:10.4018/978-1-5225-7071-4.ch003

Briggs, J. R. (2018). Immersive Wearables: Their Political and Social Effects and What Both Mean for Western Liberal Democracy. *International Journal of Sociotechnology and Knowledge Development, 10*(3), 54–71. doi:10.4018/IJSKD.2018070104

Brunet, T., & Ramachandran, P. G. (2017). Accessible and Inclusive Content and Applications. In S. Mukherjea (Ed.), *Mobile Application Development, Usability, and Security* (pp. 54–67). Hershey, PA: IGI Global. doi:10.4018/978-1-5225-0945-5.ch003

Chakraborty, C. (2019). Mobile Health (M-Health) for Tele-Wound Monitoring: Role of M-Health in Wound Management. In A. Moumtzoglou (Ed.), *Mobile Health Applications for Quality Healthcare Delivery* (pp. 98–116). Hershey, PA: IGI Global. doi:10.4018/978-1-5225-8021-8.ch005

Chen, E. T. (2018). The Impact of Healthcare Information Technology on Patient Outcomes. *International Journal of Public Health Management and Ethics, 3*(2), 39–56. doi:10.4018/IJPHME.2018070103

Corno, F., De Russis, L., & Roffarello, A. M. (2017). IoT for Ambient Assisted Living: Care4Me – A Healthcare Support System. In C. Reis & M. Maximiano (Eds.), *Internet of Things and Advanced Application in Healthcare* (pp. 66–97). Hershey, PA: IGI Global. doi:10.4018/978-1-5225-1820-4.ch003

D'Angelo, T., Delabrida Silva, S. E., Oliveira, R. A., & Loureiro, A. A. (2018). Development of a Low-Cost Augmented Reality Head-Mounted Display Prototype. In S. Delabrida Silva, R. Rabelo Oliveira, & A. Loureiro (Eds.), *Examining Developments and Applications of Wearable Devices in Modern Society* (pp. 1–28). Hershey, PA: IGI Global. doi:10.4018/978-1-5225-3290-3.ch001

da Costa, F., & de Sá-Soares, F. (2017). Authenticity Challenges of Wearable Technologies. In A. Marrington, D. Kerr, & J. Gammack (Eds.), *Managing Security Issues and the Hidden Dangers of Wearable Technologies* (pp. 98–130). Hershey, PA: IGI Global. doi:10.4018/978-1-5225-1016-1.ch005

Das, P. K., Ghosh, D., Jagtap, P., Joshi, A., & Finin, T. (2017). Preserving User Privacy and Security in Context-Aware Mobile Platforms. In S. Mukherjea (Ed.), *Mobile Application Development, Usability, and Security* (pp. 166–193). Hershey, PA: IGI Global. doi:10.4018/978-1-5225-0945-5.ch008

Domingos, D., Respício, A., & Martinho, R. (2017). Reliability of IoT-Aware BPMN Healthcare Processes. In C. Reis & M. Maximiano (Eds.), *Internet of Things and Advanced Application in Healthcare* (pp. 214–248). Hershey, PA: IGI Global. doi:10.4018/978-1-5225-1820-4.ch008

Ekambaram, V., Sharma, V., & Rajput, N. (2017). Mobile Application Testing. In S. Mukherjea (Ed.), *Mobile Application Development, Usability, and Security* (pp. 25–53). Hershey, PA: IGI Global. doi:10.4018/978-1-5225-0945-5.ch002

Evans, K. S., & Wang, E. B. (2019). Data Analysis and Integration in Healthcare. In N. Bouchemal (Ed.), *Intelligent Systems for Healthcare Management and Delivery* (pp. 220–234). Hershey, PA: IGI Global. doi:10.4018/978-1-5225-7071-4.ch010

Gammack, J., & Marrington, A. (2017). The Promise and Perils of Wearable Technologies. In A. Marrington, D. Kerr, & J. Gammack (Eds.), *Managing Security Issues and the Hidden Dangers of Wearable Technologies* (pp. 1–17). Hershey, PA: IGI Global. doi:10.4018/978-1-5225-1016-1.ch001

Gautam, P., Sunkaria, R. K., & Sharma, L. D. (2019). Digitization of Paper Electrocardiogram: A Review. In D. Kisku, P. Gupta, & J. Sing (Eds.), Design and Implementation of Healthcare Biometric Systems (pp. 212-228). Hershey, PA: IGI Global. doi:10.4018/978-1-5225-7525-2.ch009

Gavrilova, M. L., Ahmed, F., Bari, A. S., Liu, R., Liu, T., Maret, Y., . . . Sudhakar, T. (2019). Multi-Modal Motion-Capture-Based Biometric Systems for Emergency Response and Patient Rehabilitation. In D. Kisku, P. Gupta, & J. Sing (Eds.), Design and Implementation of Healthcare Biometric Systems (pp. 160-184). Hershey, PA: IGI Global. doi:10.4018/978-1-5225-7525-2.ch007

Govinda, K. (2018). IoT in the Field of Healthcare. In R. Rajkumar (Ed.), *Contemporary Applications of Mobile Computing in Healthcare Settings* (pp. 1–20). Hershey, PA: IGI Global. doi:10.4018/978-1-5225-5036-5.ch001

Govinda, K. (2018). Geo-Location-Based File Security System for Healthcare Data. In R. Rajkumar (Ed.), *Contemporary Applications of Mobile Computing in Healthcare Settings* (pp. 125–135). Hershey, PA: IGI Global. doi:10.4018/978-1-5225-5036-5.ch007

Govinda, K. (2018). Clinical Data Analysis Using IoT Devices. In R. Rajkumar (Ed.), *Contemporary Applications of Mobile Computing in Healthcare Settings* (pp. 136–153). Hershey, PA: IGI Global. doi:10.4018/978-1-5225-5036-5.ch008

Govinda, K. (2018). Body Fitness Monitoring Using IoT Device. In R. Rajkumar (Ed.), *Contemporary Applications of Mobile Computing in Healthcare Settings* (pp. 154–169). Hershey, PA: IGI Global. doi:10.4018/978-1-5225-5036-5.ch009

Govinda, K., & Ramasubbareddy, S. (2018). Smart Healthcare Administration Over Cloud. In R. Rajkumar (Ed.), *Contemporary Applications of Mobile Computing in Healthcare Settings* (pp. 34–50). Hershey, PA: IGI Global. doi:10.4018/978-1-5225-5036-5.ch003

Hao, B., & Hei, X. (2019). Voice Liveness Detection for Medical Devices. In D. Kisku, P. Gupta, & J. Sing (Eds.), Design and Implementation of Healthcare Biometric Systems (pp. 109-136). Hershey, PA: IGI Global. doi:10.4018/978-1-5225-7525-2.ch005

Hariharan, R. (2018). Wearable Internet of Things. In S. Delabrida Silva, R. Rabelo Oliveira, & A. Loureiro (Eds.), *Examining Developments and Applications of Wearable Devices in Modern Society* (pp. 29–57). Hershey, PA: IGI Global. doi:10.4018/978-1-5225-3290-3.ch002

Kamboj, A., Rani, R., & Nigam, A. (2019). EarLocalizer: A Deep-Learning-Based Ear Localization Model for Side Face Images in the Wild. In D. Kisku, P. Gupta, & J. Sing (Eds.), Design and Implementation of Healthcare Biometric Systems (pp. 137-159). Hershey, PA: IGI Global. doi:10.4018/978-1-5225-7525-2.ch006

Karthick, G. S., & Pankajavalli, P. B. (2019). Healthcare IoT Architectures, Technologies, Applications, and Issues: A Deep Insight. In N. Bouchemal (Ed.), *Intelligent Systems for Healthcare Management and Delivery* (pp. 235–265). Hershey, PA: IGI Global. doi:10.4018/978-1-5225-7071-4.ch011

Kashyap, R. (2019). Security, Reliability, and Performance Assessment for Healthcare Biometrics. In D. Kisku, P. Gupta, & J. Sing (Eds.), Design and Implementation of Healthcare Biometric Systems (pp. 29-54). Hershey, PA: IGI Global. doi:10.4018/978-1-5225-7525-2.ch002

Kasiviswanathan, U., Kushwaha, A., & Sharma, S. (2019). Development of Human Speech Signal-Based Intelligent Human-Computer Interface for Driving a Wheelchair in Enhancing the Quality-of-Life of the Persons. In N. Bouchemal (Ed.), *Intelligent Systems for Healthcare Management and Delivery* (pp. 21–60). Hershey, PA: IGI Global. doi:10.4018/978-1-5225-7071-4.ch002

Kerr, D., Butler-Henderson, K., & Sahama, T. (2017). Security, Privacy, and Ownership Issues with the Use of Wearable Health Technologies. In A. Marrington, D. Kerr, & J. Gammack (Eds.), *Managing Security Issues and the Hidden Dangers of Wearable Technologies* (pp. 161–181). Hershey, PA: IGI Global. doi:10.4018/978-1-5225-1016-1.ch007

Kerr, D., & Gammack, J. (2017). Conclusions: Where Next for Wearables? In A. Marrington, D. Kerr, & J. Gammack (Eds.), *Managing Security Issues and the Hidden Dangers of Wearable Technologies* (pp. 289–299). Hershey, PA: IGI Global. doi:10.4018/978-1-5225-1016-1.ch012

Khalifa, S., Lan, G., Hassan, M., Hu, W., & Seneviratne, A. (2018). Human Context Detection From Kinetic Energy Harvesting Wearables. In S. Delabrida Silva, R. Rabelo Oliveira, & A. Loureiro (Eds.), *Examining Developments and Applications of Wearable Devices in Modern Society* (pp. 107–133). Hershey, PA: IGI Global. doi:10.4018/978-1-5225-3290-3.ch005

Koumpouros, Y., & Georgoulas, A. (2019). The Rise of mHealth Research in Europe: A Macroscopic Analysis of EC-Funded Projects of the Last Decade. In A. Moumtzoglou (Ed.), *Mobile Health Applications for Quality Healthcare Delivery* (pp. 1–29). Hershey, PA: IGI Global. doi:10.4018/978-1-5225-8021-8.ch001

Kouroubali, A., Koumakis, L., Kondylakis, H., & Katehakis, D. G. (2019). An Integrated Approach Towards Developing Quality Mobile Health Apps for Cancer. In A. Moumtzoglou (Ed.), *Mobile Health Applications for Quality Healthcare Delivery* (pp. 46–71). Hershey, PA: IGI Global. doi:10.4018/978-1-5225-8021-8.ch003

Kumar, U., Tripathi, E., Tripathi, S. P., & Gupta, K. K. (2019). Deep Learning for Healthcare Biometrics. In D. Kisku, P. Gupta, & J. Sing (Eds.), Design and Implementation of Healthcare Biometric Systems (pp. 73-108). Hershey, PA: IGI Global. doi:10.4018/978-1-5225-7525-2.ch004

Lee, H. (2017). The Internet of Things and Assistive Technologies for People with Disabilities: Applications, Trends, and Issues. In C. Reis & M. Maximiano (Eds.), *Internet of Things and Advanced Application in Healthcare* (pp. 32–65). Hershey, PA: IGI Global. doi:10.4018/978-1-5225-1820-4.ch002

Manoj, A. S., Hussain, M. A., & Teja, P. S. (2019). Patient Health Monitoring Using IoT. In A. Moumtzoglou (Ed.), *Mobile Health Applications for Quality Healthcare Delivery* (pp. 30–45). Hershey, PA: IGI Global. doi:10.4018/978-1-5225-8021-8.ch002

Metelmann, B., & Metelmann, C. (2019). Mobile Health Applications in Prehospital Emergency Medicine. In A. Moumtzoglou (Ed.), *Mobile Health Applications for Quality Healthcare Delivery* (pp. 117–135). Hershey, PA: IGI Global. doi:10.4018/978-1-5225-8021-8.ch006

Michael, K., Gokyer, D., & Abbas, S. (2017). Societal Implications of Wearable Technology: Interpreting “Trialability on the Run”. In A. Marrington, D. Kerr, & J. Gammack (Eds.), *Managing Security Issues and the Hidden Dangers of Wearable Technologies* (pp. 238–266). Hershey, PA: IGI Global. doi:10.4018/978-1-5225-1016-1.ch010

Mishra, S., & Panda, M. (2019). Artificial Intelligence in Medical Science. In N. Bouchemal (Ed.), *Intelligent Systems for Healthcare Management and Delivery* (pp. 306–330). Hershey, PA: IGI Global. doi:10.4018/978-1-5225-7071-4.ch014

Moumtzoglou, A. S. (2019). The Science of Individuality and Tailored M-Health Communication. In A. Moumtzoglou (Ed.), *Mobile Health Applications for Quality Healthcare Delivery* (pp. 213–234). Hershey, PA: IGI Global. doi:10.4018/978-1-5225-8021-8.ch010

Moumtzoglou, A. S. (2019). The Nexus of M-Health and Self-Care. In A. Moumtzoglou (Ed.), *Mobile Health Applications for Quality Healthcare Delivery* (pp. 235–259). Hershey, PA: IGI Global. doi:10.4018/978-1-5225-8021-8.ch011

Mukherjee, T., Chander, D., Eswaran, S., & Dasgupta, K. (2017). Participatory Sensing for City-Scale Applications. In S. Mukherjea (Ed.), *Mobile Application Development, Usability, and Security* (pp. 210–230). Hershey, PA: IGI Global. doi:10.4018/978-1-5225-0945-5.ch010

Naaz, S., & Siddiqui, F. (2019). Application of Big Data in Digital Epidemiology. In N. Bouchemal (Ed.), *Intelligent Systems for Healthcare Management and Delivery* (pp. 285–305). Hershey, PA: IGI Global. doi:10.4018/978-1-5225-7071-4.ch013

Nadler, S. (2017). Mobile Location Tracking: Indoor and Outdoor Location Tracking. In S. Mukherjea (Ed.), *Mobile Application Development, Usability, and Security* (pp. 194–209). Hershey, PA: IGI Global. doi:10.4018/978-1-5225-0945-5.ch009

Narayana Moorthi, M., & Manjula, R. (2018). A Survey of Mobile Computing Devices and Sensors in Healthcare Applications: Real-Time System Design. In R. Rajkumar (Ed.), *Contemporary Applications of Mobile Computing in Healthcare Settings* (pp. 51–57). Hershey, PA: IGI Global. doi:10.4018/978-1-5225-5036-5.ch004

Narváez, P., Manjarrés, J., Percybrooks, W., Pardo, M., & Calle, M. (2019). Assessing the Level of Physical Activity in the Workplace: A Case Study With Wearable Technology. *International Journal of Interdisciplinary Telecommunications and Networking, 11*(1), 44–56. doi:10.4018/IJITN.2019010104

Neto, J. S., Silva, A. L., Nakano, F., Pérez-Álcazar, J. J., & Kofuji, S. T. (2018). When Wearable Computing Meets Smart Cities: Assistive Technology Empowering Persons With Disabilities. In S. Delabrida Silva, R. Rabelo Oliveira, & A. Loureiro (Eds.), *Examining Developments and Applications of Wearable Devices in Modern Society* (pp. 58–85). Hershey, PA: IGI Global. doi:10.4018/978-1-5225-3290-3.ch003

Nimkar, S., & Gilles, E. E. (2018). Improving Global Health With Smartphone Technology: A Decade in Review of mHealth Initiatives. *International Journal of E-Health and Medical Communications*, *9*(3), 1–19. doi:10.4018/IJEHMC.2018070101

Pistoia, M., Tripp, O., & Lubensky, D. (2017). Combining Static Code Analysis and Machine Learning for Automatic Detection of Security Vulnerabilities in Mobile Apps. In S. Mukherjea (Ed.), *Mobile Application Development, Usability, and Security* (pp. 68–94). Hershey, PA: IGI Global. doi:10.4018/978-1-5225-0945-5.ch004

Pouliakis, A., Karakitsou, E., & Margari, N. (2019). Cytopathology and the Smartphone: An Update. In A. Moumtzoglou (Ed.), *Mobile Health Applications for Quality Healthcare Delivery* (pp. 136–164). Hershey, PA: IGI Global. doi:10.4018/978-1-5225-8021-8.ch007

Pramanik, P. K., Pal, S., & Mukhopadhyay, M. (2019). Healthcare Big Data: A Comprehensive Overview. In N. Bouchemal (Ed.), *Intelligent Systems for Healthcare Management and Delivery* (pp. 72–100). Hershey, PA: IGI Global. doi:10.4018/978-1-5225-7071-4.ch004

Rajkumar, R. (2018). Smart Healthcare. In R. Rajkumar (Ed.), *Contemporary Applications of Mobile Computing in Healthcare Settings* (pp. 21–33). Hershey, PA: IGI Global. doi:10.4018/978-1-5225-5036-5.ch002

Rakshit, R. D., & Kisku, D. R. (2019). Biometric Technologies in Healthcare Biometrics. In D. Kisku, P. Gupta, & J. Sing (Eds.), Design and Implementation of Healthcare Biometric Systems (pp. 1-28). Hershey, PA: IGI Global. doi:10.4018/978-1-5225-7525-2.ch001

Ramakrishna, V., & Dey, K. (2017). Mobile Application and User Analytics. In S. Mukherjea (Ed.), *Mobile Application Development, Usability, and Security* (pp. 231–259). Hershey, PA: IGI Global. doi:10.4018/978-1-5225-0945-5.ch011

Ranvier, J., Catasta, M., Gavrilovic, I., Christodoulou, G., Radu, H., Signo', T., & Aberer, K. (2017). MEmoIt: From Lifelogging Application to Research Platform. In S. Mukherjea (Ed.), *Mobile Application Development, Usability, and Security* (pp. 1–24). Hershey, PA: IGI Global. doi:10.4018/978-1-5225-0945-5.ch001

Rath, M. (2018). Technical and Operational Utility of Ubiquitous Devices with Challenging Issues in Emerging Ubiquitous Computing. *International Journal of Mobile Devices, Wearable Technology, and Flexible Electronics*, *9*(1), 16–35. doi:10.4018/IJMDWTFE.2018010102

Resnick, M. L., & Chircu, A. M. (2017). Wearable Devices: Ethical Challenges and Solutions. In A. Marrington, D. Kerr, & J. Gammack (Eds.), *Managing Security Issues and the Hidden Dangers of Wearable Technologies* (pp. 182–205). Hershey, PA: IGI Global. doi:10.4018/978-1-5225-1016-1.ch008

Ricci, J., Baggili, I., & Breitinger, F. (2017). Watch What You Wear: Smartwatches and Sluggish Security. In A. Marrington, D. Kerr, & J. Gammack (Eds.), *Managing Security Issues and the Hidden Dangers of Wearable Technologies* (pp. 47–73). Hershey, PA: IGI Global. doi:10.4018/978-1-5225-1016-1.ch003

Sabrina, Y., & Tayeb, L. M. (2019). Edge Detection on Light Field Images: Evaluation of Retinal Blood Vessels Detection on a Simulated Light Field Fundus Photography. In N. Bouchemal (Ed.), *Intelligent Systems for Healthcare Management and Delivery* (pp. 174–197). Hershey, PA: IGI Global. doi:10.4018/978-1-5225-7071-4.ch007

Sánchez, Y. K., Demurjian, S. A., Conover, J., Agresta, T. P., Shao, X., & Diamond, M. (2017). Role-Based Access Control for Mobile Computing and Applications. In S. Mukherjea (Ed.), *Mobile Application Development, Usability, and Security* (pp. 117–141). Hershey, PA: IGI Global. doi:10.4018/978-1-5225-0945-5.ch006

Sanzi, E., Demurjian, S. A., Agresta, T. P., & Murphy, A. (2017). Trust Profiling to Enable Adaptive Trust Negotiation in Mobile Devices. In S. Mukherjea (Ed.), *Mobile Application Development, Usability, and Security* (pp. 95–116). Hershey, PA: IGI Global. doi:10.4018/978-1-5225-0945-5.ch005

Sasikala, R., & Sureshkumar, N. (2019). Research Investigation and Analysis on Behavioral Analytics, Neuro Imaging, and Pervasive Sensory Algorithms and Techniques for Autism Diagnosis. In N. Bouchemal (Ed.), *Intelligent Systems for Healthcare Management and Delivery* (pp. 206–219). Hershey, PA: IGI Global. doi:10.4018/978-1-5225-7071-4.ch009

Shao, X., Demurjian, S. A., & Agresta, T. P. (2017). A Spatio-Situation-Based Access Control Model for Dynamic Permission on Mobile Applications. In S. Mukherjea (Ed.), *Mobile Application Development, Usability, and Security* (pp. 142–165). Hershey, PA: IGI Global. doi:10.4018/978-1-5225-0945-5.ch007

Silva, A. D., Rigo, S. J., & Barbosa, J. L. (2018). Wearable Health Care Ubiquitous System for Stroke Monitoring and Alert. In S. Delabrida Silva, R. Rabelo Oliveira, & A. Loureiro (Eds.), *Examining Developments and Applications of Wearable Devices in Modern Society* (pp. 134–160). Hershey, PA: IGI Global. doi:10.4018/978-1-5225-3290-3.ch006

Singh, P. (2017). Mobile + Cloud: Opportunities and Challenges. In S. Mukherjea (Ed.), *Mobile Application Development, Usability, and Security* (pp. 260–279). Hershey, PA: IGI Global. doi:10.4018/978-1-5225-0945-5.ch012

Sinha, N., & Gupta, M. (2019). Taxonomy of Wearable Devices: A Systematic Review of Literature. *International Journal of Technology Diffusion*, *10*(2), 1–17. doi:10.4018/IJTD.2019040101

Skapura, N., & Dong, G. (2017). Class Distribution Curve Based Discretization With Application to Wearable Sensors and Medical Monitoring. *International Journal of Monitoring and Surveillance Technologies Research*, *5*(4), 23–37. doi:10.4018/IJMSTR.2017100102

Spaanenburg, L. (2017). The Role of Time in Health IoT. In C. Reis & M. Maximiano (Eds.), *Internet of Things and Advanced Application in Healthcare* (pp. 197–213). Hershey, PA: IGI Global. doi:10.4018/978-1-5225-1820-4.ch007

Srinivasa, K. G., Sowmya, B. J., Shikhar, A., Utkarsha, R., & Singh, A. (2018). Data Analytics Assisted Internet of Things Towards Building Intelligent Healthcare Monitoring Systems: IoT for Healthcare. *Journal of Organizational and End User Computing*, *30*(4), 83–103. doi:10.4018/JOEUC.2018100106

Sruthi, M., & Rajasekaran, R. (2018). Reading Assistance for Visually Impaired People Using TTL Serial Camera With Voice. In R. Rajkumar (Ed.), *Contemporary Applications of Mobile Computing in Healthcare Settings* (pp. 170–180). Hershey, PA: IGI Global. doi:10.4018/978-1-5225-5036-5.ch010

Sultana, H. P., & Nagendran, N. (2018). Mobile Patient Surveillance. In R. Rajkumar (Ed.), *Contemporary Applications of Mobile Computing in Healthcare Settings* (pp. 58–84). Hershey, PA: IGI Global. doi:10.4018/978-1-5225-5036-5.ch005

Tiwari, K., Arora, G., & Gupta, P. (2019). Indexing for Healthcare Biometric Databases. In D. Kisku, P. Gupta, & J. Sing (Eds.), Design and Implementation of Healthcare Biometric Systems (pp. 55-72). Hershey, PA: IGI Global. doi:10.4018/978-1-5225-7525-2.ch003

Trauth, E., & Browning, E. R. (2018). Technologized Talk: Wearable Technologies, Patient Agency, and Medical Communication in Healthcare Settings. *International Journal of Sociotechnology and Knowledge Development, 10*(3), 1–26. doi:10.4018/IJSKD.2018070101

Turcu, C. E., & Turcu, C. O. (2017). Social Internet of Things in Healthcare: From Things to Social Things in Internet of Things. In C. Reis & M. Maximiano (Eds.), *Internet of Things and Advanced Application in Healthcare* (pp. 266–295). Hershey, PA: IGI Global. doi:10.4018/978-1-5225-1820-4.ch010

Ünal, B. C., & Ünal, O. (2017). Wind Turbine Remote Maintenance With Wearable Technologies. *International Journal of Green Computing, 8*(1), 36–54. doi:10.4018/IJGC.2017010103

Vasanthamani, S. (2018). A Study on Lifetime Enhancement and Reliability in Wearable Wireless Body Area Networks. *International Journal of User-Driven Healthcare, 8*(2), 46–59. doi:10.4018/IJUDH.2018070103

Venugopal, M. (2019). Evolution of Digital Technologies and Use of Virtual Assistants in Drug Development. In N. Bouchemal (Ed.), *Intelligent Systems for Healthcare Management and Delivery* (pp. 1–20). Hershey, PA: IGI Global. doi:10.4018/978-1-5225-7071-4.ch001

Vlachopapadopoulou, E., Georga, E. I., & Fotiadis, D. I. (2019). Management of Obese Pediatric Patients in the Digital Era. In A. Moumtzoglou (Ed.), *Mobile Health Applications for Quality Healthcare Delivery* (pp. 72–97). Hershey, PA: IGI Global. doi:10.4018/978-1-5225-8021-8.ch004

Wadhera, T., Kakkar, D., Kaur, G., & Menia, V. (2019). Pre-Clinical ASD Screening Using Multi-Biometrics-Based Systems. In D. Kisku, P. Gupta, & J. Sing (Eds.), Design and Implementation of Healthcare Biometric Systems (pp. 185-211). Hershey, PA: IGI Global. doi:10.4018/978-1-5225-7525-2.ch008

Wang, M., & Kerr, D. (2017). Confidential Data Storage Systems for Wearable Platforms. In A. Marrington, D. Kerr, & J. Gammack (Eds.), *Managing Security Issues and the Hidden Dangers of Wearable Technologies* (pp. 74–97). Hershey, PA: IGI Global. doi:10.4018/978-1-5225-1016-1.ch004

Xie, L., Zheng, L., & Yang, G. (2017). Hybrid Integration Technology for Wearable Sensor Systems. In C. Reis & M. Maximiano (Eds.), *Internet of Things and Advanced Application in Healthcare* (pp. 98–137). Hershey, PA: IGI Global. doi:10.4018/978-1-5225-1820-4.ch004

Yellampalli, S. S., Kiran, N. R., & Malapur, I. (2019). Medi-Rings for Senior Citizens: Distributed EMR System. In N. Bouchemal (Ed.), *Intelligent Systems for Healthcare Management and Delivery* (pp. 148–173). Hershey, PA: IGI Global. doi:10.4018/978-1-5225-7071-4.ch006

Yucesan, M., Gul, M., Mete, S., & Celik, E. (2019). A Forecasting Model for Patient Arrivals of an Emergency Department in Healthcare Management Systems. In N. Bouchemal (Ed.), *Intelligent Systems for Healthcare Management and Delivery* (pp. 266–284). Hershey, PA: IGI Global. doi:10.4018/978-1-5225-7071-4.ch012

Zayed, N. M., & Elnemr, H. A. (2019). Deep Learning and Medical Imaging. In N. Bouchemal (Ed.), *Intelligent Systems for Healthcare Management and Delivery* (pp. 101–147). Hershey, PA: IGI Global. doi:10.4018/978-1-5225-7071-4.ch005

Zimeras, S. (2019). Mathematical Models for Computer Virus: Computer Virus Epidemiology. In A. Moumtzoglou (Ed.), *Mobile Health Applications for Quality Healthcare Delivery* (pp. 203–212). Hershey, PA: IGI Global. doi:10.4018/978-1-5225-8021-8.ch009

About the Contributors

Vinod Kumar Singh received his Ph.D. degree in Electronics & Communication Engineering and M. Tech. in Digital Communication System in 2013 & 2009 respectively. He is having about Sixteen years of experience in the field of Electrical and Electronics Engineering. He is a Professor and Head in Electrical Engineering Department of S.R. Group of Institutions, Jhansi UP, India. He is a senior member of International Association of Computer Science and Information Technology (IACSIT) and International Association for the engineers & computer scientists (IAENG). Prof. Singh has published more than 200 research papers in the renowned International Journals such as IEEE, Springer and Willey.

Zakir Ali received his Ph.D. degree in Electronics & Communication Engineering from B.U. Rajasthan, India in 2013. He received M. Tech. in Digital Communication System from B.I.E.T., Jhansi, India in 2009. He is having about Nine years of experience in the field of Electronics & Communication Engineering. He is an Assistant Professor in Electronics & Communication Engineering Department of IET Bundelkhand University, Jhansi at UP, India. He is a senior member of International Association of Computer Science and Information Technology (IACSIT) and International Association for the engineers & computer scientists (IAENG). Dr. Ali has published more than 40 research papers in the renowned International Journals such as Springer, Elsevier. His research interests include design of wide and dual band microstrip antenna, Ultra wide band textile antenna. He is a Reviewer of many SCI International Journals.

* * *

Nikhil Agrawal received B.tech degree with honors in Electrical Engineering in 2014 and M.E. in Industrial systems & Drive in 2017. He awarded with M.I.T.S. Gold Medal 2017 for securing first position in M.E. (Industrial systems& Drives). After M.E. he joined the Ujjain engineering College, Ujjain as a lecturer, recently Mr. Agrawal join S.R. group of institutions, Jhansi as an Assistant Professor. He has published research papers in IEEE conferences, International journal and national conferences.

Naresh B. received B.Tech and M.Tech degrees in Electrical and Electronics Engineering from Jawaharlal Nehru Technological University, Hyderabad, India in 2006 and 2012. He is currently pursuing Ph.D in Electrical Engineering from Bhagewant University, Ajmer, Rajasthan, India.

Priyanka Bansal received the Bachelor's degree in Electronics & Communication Engineering from Vaish College of Engineering, Rohtak and M. Tech degree (Electronics & Communication Engineering) from Career Institute of Technology & Management, Faridabad both affiliated to Maharishi Dayanand University, Rohtak and Ph.D. degree in Electronics & Communication Engineering from Jamia Millia Islamia (A Central University), New Delhi. Since 2003, she is indulged in teaching at engineering colleges. Presently she is working as Associate Professor in the Department of Electronics and Communication Engineering at Manav Rachna International Institute of Research & Studies, Faridabad. Her research interests include Signal Processing and Antenna Designed. She has published 10 research papers in journals of repute and 20 research papers in National/International conferences. She has guided more than 20 undergraduate projects and 5 postgraduate projects till date. She is a life member of Institution of Engineers.

Rishabh Kumar Baudh was born in Chirgaon, Uttar Pradesh, in 1992. He received the B.Tech degree in Electronics and Communication from Uttar Pradesh Technical University, Lucknow, in 2014. He is currently working towards the M.Tech degree in Bundelkhand Institute of Engineering and Technology (BIET), Jhansi, from Dr. APJ Abdul Kalam Technical University(AKTU), Lucknow, India. He has worked as Lecturer at SR Group of Institution, Jhansi, U.P, India from Sep. 2015 to July 2017. His current research area is textile antenna, RF Energy Harvesting and microstrip antenna theory and design.

Akash Kumar Bhoi is Assistant Professor in Department of Electrical & Electronics Engineering, Sikkim Manipal Institute of Technology, Sikkim Manipal University, Sikkim, India.

Devi Charan Dhubkarya received the M.Tech degree in microwave and radar engineering from Indian Institute of Technology(IIT), Roorkee. He completed his Ph.D in Microwave engineering from Bundelkhand University, Jhansi, Uttar Pradesh. Currently he is working as a Associate Professor in Electronics and Communication department of Bundelkhand Institute of Engineering and Technology, Jhansi, from Dr. APJ Abdul Kalam Technical University(AKTU), Lucknow, India. He worked as a Scientist in microwave tubes domain in CEERI Pilani, Rajasthan. He has 27 years of experience in research and teaching. He has published more than 70 research papers in his research area of microwave antenna, wireless communication, textile antenna, etc.

Nupur Gupta was born in Jhansi, Uttar Pradesh, in 1991. She received the Bachelors of engineering degree in Electronics and Communication from Rajiv Gandhi Proudyogiki Vishwavidyalaya, Bhopal, Madhya Pradesh, in 2014. She received the M.Tech degree from Bundelkhand Institute of Engineering and Technology (BIET), Jhansi, from Dr. APJ Abdul Kalam Technical University(AKTU), Lucknow, India in 2018. Her research work area is concerned in textile antenna, microstrip antennas for wireless communication.

Sudesh Gupta is working as Assistant Professor in the department of Electronics & Communication Engineering. He is having more than 18 years of experience in the field of teaching. His area of specialization is "Digital Communication, Digital Circuits & System etc.". He has published – 4 research papers in National and International Journals. He has actively attended Faculty Development Programs, seminars and participated in National and International conferences, organized by reputed organizations.

Swati Jain pursuing Master in Engineering in Power system from University Institute of Technology RGPV Bhopal and completed Bachelor of Technology in Electrical and Electronics Engineering from Sagar Institute of Research and Technology Bhopal.

Bharat Bhushan Khare completed a bachelor of technology with honors from SRGI Jhansi, India in Electrical Engineering and is now pursuing Master of Engineering in Power System from UIT RGPV Bhopal, India. Mr. Khare already published some research papers on RF Energy Harvesting topics in international journals and conferences and is trying to increase the usage of wireless power transmission by using antenna array with rectifier circuit.

Ravi Kant Prasad was born in Munger, Bihar, in 1982. He received the B.Tech degree in Electronics and Telecommunication and the M.Tech degree in Digital Communication from Uttarpradesh Technical University, Lucknow, in 2005 and 2013. He is currently working towards the Ph.D degree in Bundelkhand Institute of Engineering and Technology (BIET), Jhansi, Uttarpradesh Technical University, Lucknow, India. He has worked as Lecturer at BBIET & RC, Bulandshahar, U.P, India from Aug. 2005 to July 2008 and as Assistant Professor, IIMT College of Engineering, Greater Noida, U.P, India from Aug. 2008 to June 2014. His current research area is microstrip antenna theory and design.

Anurag Saxena received his Ph.D. degree in Electronics & Communication Engineering from C.S.J.M., India in 2016. He received M. Tech. in Digital Communication System from Manav Bharti University, Solan, India in 2013. He is having about Eight years of experience in the field of Electronics Engineering. He is an Assistant Professor in Electrical Engineering Department of S.R. Group of Institutions, Jhansi at UP, India. Dr. Saxena has published more than 15 research papers in the renowned International Journals such as Springer. His research interests include design of wide microstrip antenna, Ultra wide band textile antenna.

Vimlesh Singh received the Bachelor's degree in Electronics & Communication Engineering from KNMIET Modinagar, the M. Tech degree (Digital & Communication Engineering) from Bundelkhand Institute of Engineering & Technology Jhansi affiliated to UP Technical University. She has been completed her Ph.D. degree in Electronics & Communication Engineering from JJTU, Rajasthan in 2016. She was lecturer in Institute of Engineering and Technology, Bundelkhand University Jhansi from 2003 to 2008. Presently she is working as Associate Professor, in Faculty of Engineering and Technology, Manav Rachan International Institute of Research and studies from 2008 to till date. Her research interests are in Microwave Antennas and VLSI design. At present, she is engaged in Fractal antenna design for wide bandwidth which has numerous application in communication Engineering. he has published several paper in national and international conferences and journals. She is life member of various technical professional society.

D. K. Srivastava is Associate Professor at the Electronics and Communication Engineering Dept. of Bundelkhand Institute of Engineering and Technology (BIET), Jhansi. He accomplished his Ph. D. in Electronics Engineering in 2010. Earlier, he obtained the M. Tech. in Electronics Engineering in 1999 and the B. Tech. in Electronics Engineering in 1994. Dr. Srivastava has published more than 120 International /National research papers and Supervised Ph.D. and M. Tech, thesis. He is reviewer of 'Progress in Electromagnetic Research' (PIER). He has long standing academic interests in microstrip antenna, digital electronics and optical communication. He has delivered many lectures/keynote addresses and chaired various technical sessions during national and international conferences and training programs.

Gaurav Varma was born in Chirgaon, Uttar Pradesh, in 1992. He received the B Tech degree in Electronics and Communication from Bundelkhand University, Jhansi, in 2017. He is currently working towards the M Tech degree in Bundelkhand Institute of Engineering and technology (BIET), Jhansi, from Dr. APJ Abdul kalam Technical University (AKTU), Lucknow, India. His current research area is textile antenna, wearable microstrip antenna theory and design.

Pranay Yadav (M'17) received the B.E. degree E.C. from RGPV (2011) and Masters (M.E.) degree in Digital Communication from RGPV (2013). Currently working with Ultra-Light technology as a project manager of R&D department, former worked as a Lecturer, Assistant professor and Head of department. His research work focuses on modern optimization technologies to industry and control processes in field of Real Time applications based on IoT, Cloud, Automation, Embedded system, Microstrip patch antenna, flexible antenna and MEMS Technology. Machine learning, optimization of system, IoT based Image Processing. His technical expertise are in image processing, computer vision and microstrip antenna designing.

Index

P

R

S

T

W